A History of Chemical Warfare

A History of Chemical Warfare

Kim Coleman

First published 2005 by
PALGRAVE MACMILLAN
Houndmills, Basingstoke, Hampshire RG21 6XS and
175 Fifth Avenue, New York, N.Y. 10010
Companies and representatives throughout the world

PALGRAVE MACMILLAN is the global academic imprint of the Palgrave
Macmillan division of St. Martin's Press, LLC and of Palgrave Macmillan Ltd.
Macmillan® is a registered trademark in the United States, United Kingdom
and other countries. Palgrave is a registered trademark in the European
Union and other countries.

ISBN-13: 978–1–4039–3459–8 hardback
ISBN-10: 1–4039–3459–2 hardback
ISBN-13: 978–1–4039–3460–4 paperback
ISBN-10: 1–4039–3460–6 paperback

This book is printed on paper suitable for recycling and made from fully
managed and sustained forest sources.

A catalogue record for this book is available from the British Library.

Library of Congress Cataloging-in-Publication Data
Coleman, Kim, 1953–
    A history of chemical warfare / Kim Coleman.
       p.   cm.
    Includes bibliographical references and index.
    ISBN 1–4039–3459–2 (cloth) — ISBN 1–4039–3460–6 (pbk.)
    1. Chemical warfare—History. I. Title.

    UG447.C637 2005
    358'.34'09—dc22

                                                     2004065796

10   9   8   7   6   5   4   3   2   1
14   13   12   11   10   09   08   07   06   05

Printed and bound in Great Britain by
Antony Rowe Ltd, Chippenham and Eastbourne

For my Dad.

A promise kept.

Rest in peace.

also

For Amy

I know it's the last day on earth, We'll be together while the planet dies,
I know it's the last day on earth, We'll never say Goodbye

– M. Manson

Contents

List of Tables

List of Abbreviations

BAL	British Anti-Lewisite
BEF	British Expeditionary Force
BNP	British National Party
BWC	Biological Warfare Convention
CIA	Central Intelligence Agency
CWC	Chemical Warfare Convention
DOD	Department of Defense
GWS	Gulf War Syndrome
IADL	International Association of Democratic Lawyers
IAEA	International Atomic Energy Agency
ICI	Imperial Chemical Industries Limited
IRA	Irish Republican Army
MCS	Multiple Chemistry Sensitivity
MOD	Ministry of Defence
NAPS	Nerve Agent Pre-Treatment
NATO	North Atlantic Treaty Organisation
OPCW	Organisation for the Prohibition of Chemical Weapons
PCF	French Communist Party
RAF	Royal Air Force
SIPRI	Stockholm International Peace Research Institute
SOG	Special Operations Group
TASS	Soviet News Agency
TCPs	Toxic Chemical Precursors
TWC	Toxic Weapons Convention
UNMOVIC	United Nations Monitoring, Verification and Inspection Commission
UNSCR	United Nations Security Council Resolution
UNSCOM	United Nations Special Commission
USAAF	United States Army Air Force
WMD	Weapons of Mass Destruction

List of Chemical Agents

AC	Hydrogen cyanide
BBC	Brombenzl cyanide
BZ	3-Quinuclidnyl benzilate (psychochemical gas)
CG	Phosgene (also known as carbonyl chloride and chloroformyl chloride)
CK	Cyanogen chloride
CN	Chloroacetophene (non-lethal, fast acting riot control tear gas)
CNS	CN + Chloropicrin (tear gas)
CS	Ortho-chlorobenzylidene malononitrile (toxic tear gas)
DA	Diphenylchloroarsine (tear gas)
DC	Diphenylcyanoarsine (tear gas)
DDT	Dichlorodiphenyltrichloroethane (insecticide)
DFP	Disopropylgluorophospate (nerve gas, less deadly than GA, GB or GD)
DM	Diphenylamine chlorarsine (arsenical gas, also known as Adamsite (phenarsazine chloride))
DMMP	Dimethyl methanephosphonate (water contaminant)
GA	Tabun (also known as Trilon 83, nerve agent)
GB	Sarin (nerve agent)
GD	Soman (nerve agent)
GF	Cyclosarin (nerve agent)
HD	Dichlorodiethylsulfide (mustard gas)
HN	Nitrogen mustard (vesicant)
KSK	Tear gas
Lewisite	Toxic gas
LSD (LSD-25)	Lysergic acid diethylamide
VX	Highly toxic nerve gas

Acknowledgements

In the course of the research and writing up of this book, I have run up a number of debts. My contacts with archivists in several different archives have frequently been rewarding but special thanks must go to Herr Hermann Pogarell of the Bayerarchiv, Leverkusen in Germany, Dick Bright from the Imperial War Museum Archive, Duxford and Julian Perry Robinson and Caitriona McLeish at the Science Policy Research Unit (SPRU) at the University of Sussex. Also, I would like to thank all my students at the University of Essex, especially the final-year graduates of 2004, for their unstinting support.

My greatest debt is to the one person who read and commented on all of the outlines and the entire manuscript – For this, my sincere thanks to Jon Menhinick. The fact I did not always take his good advice absolves him of the responsibility for the faults that remain in the final draft.

During the course of this research I was fortunate enough to interview several people closely associated with the events described. First, and foremost, my special thanks to Tom Jackson, ex-101st Airborne, 2nd/502nd Co. B, 'Strike Force', US Army, who served in Vietnam from 1970 to 1971 and who provided me with eyewitness accounts of napalm attacks. During the course of this research Tom became a much valued friend – thank you Tom for everything. Secondly, many thanks to Julian Perry Robinson who welcomed me to the SPRU at the University of Sussex in the latter stages of the book and provided clarification of some important aspects of the Chemical Warfare Convention. Many thanks to Luciania O'Flaherty and Daniel Bunyard at Palgrave Macmillan for guiding me through the publishing minefield and for their constant support during the preparation of this book.

Finally, I would like to acknowledge special personal debts to several persons: to my Mum who has been a constant source of support and encouragement throughout my academic career; to my sister, Jan, for her support, especially this past year; to Natalie Ashton who taught me to use a computer effectively! To my sons, James and Glenn, and my daughter, Amy, who supported and encouraged me throughout the research and writing. I owe them the most. James, thank you for your astute comments; Glenn, thank you for the 'Butterfly Effect'; and Amy, thank you for living with me through the entire process of turning an idea into a book. James, Glenn and Amy, your company and good advice have made the process very enjoyable. Thank You!

Finally, special thanks to my husband, Alan, who in an unexpected way ensured this book was completed.

Kim Coleman
London, September 2004

Preface

The twentieth century saw an unprecedented increase in destruction caused by warfare, mainly brought about by the ever-increasing lethality of weapons and the introduction of new forms of killing, notably the development and refinement of chemical weapons. The terrorist attacks on New York on 11 September 2001, and the responses to them, suggest that in the twenty-first century we shall continue to witness violence by both states and terrorist groups within them.

When compared with the methods, armaments and materials of war since the beginning of time up to the nineteenth century, the First World War marks a watershed and reveals changes which perhaps hundreds of years of peace could never have brought about. Undoubtedly, the large-scale use of chemicals as weapons came about during the First World War largely because of the unprecedented growth of the chemical industry. As the spring day of 22 April 1915 drew to a close, German soldiers released chlorine gas from cylinders against the Allies entrenched around Langemarck, near Ypres. What happened next was one of the most frightening and horrific experiences ever faced by men at war. The acrid cloud enveloped the soldiers and they began to cough, clutch their throats and gasp for air. Many turned blue and fell to the ground dead. Those who were able to escape stumbled into first-aid stations where doctors were unable to provide any effective medical treatment. Chemical warfare had begun. The feelings of shock and outrage produced by this first gas attack were compounded by the fact that poison gas was specifically outlawed by international law. The Hague Declaration of 1899, with Germany as a signatory, prohibited the use of projectiles, the object of which was the diffusion of asphyxiating gases. As horrific as they were, gas attacks were to continue for the remainder of the war. Chlorine and other agents developed by both sides claimed over 1.3 million casualties, 91,000 of whom died.[1] More than 110,000 tons of chemical agents were disseminated over the battlefields, the greater part on the Western Front. Initially the chemicals were used, not to cause casualties, in the sense of putting the enemy combatants out of action, but rather to harass. The sensory irritants used, however, were powerful enough to disable those who were exposed to them, but they served mainly to drive enemy combatants out of the trenches or other cover that protected them from conventional fire. About 10 per cent of the total tonnage of chemical warfare agents used during the First World War were chemicals of this type, consisting mainly of lacrimators (tear gases), sternutators and vomiting agents. However, the use of more lethal chemicals soon followed the introduction of disabling chemicals. Between 1915 and 1918 almost every known noxious chemical was screened in the chemical industry for its potential as a weapon and, indeed, this process was repeated during the Second World War, when substantial stocks of chemical weapons were accumulated, though never used in military operations.

Professor Fritz Haber, one of the pioneers of gas warfare, upon receiving the Nobel Prize for chemistry in 1919, made a statement that has haunted, and will continue to haunt, mankind: 'In no future war will the military be able to ignore poison gas', he predicted, 'for it is a higher form of killing'.[2] At the beginning of the twentieth century these developments, then, represented the opening of new fields and visions of possibilities previously undreamed of by the pragmatic nineteenth-century soldier. By the concentration of chemistry to war, two dominating factors emerged whose importance to war and danger to world peace gained momentum over time. The first was the scientific initiative, that is to say the invention of deadly new chemicals, and the second was the threat that their impact on war through large-scale production in the convertible industries of peace constituted. A real threat which, if unanswered in 1918 by a practical scheme for world disarmament, would aggravate the danger of a sudden, decisive attack in an otherwise disarmed world. Thus, the League of Nations, established by the Paris Peace Treaties to help attain world peace, implied recognition by our forefathers that a definite mechanism and definite measures were required. However, attempting to attain chemical peace by merely prohibiting chemical war, by expecting their mechanism to achieve its objective without functioning and to attain peace by its mere existence, is, with hindsight, naïve. Just as special measures were put in place to control the older branches of warfare, similar measures were required to control the chemical peace. As Victor Lefebure pointed out in 1919, 'Chemical peace, guaranteed by a mere signature, is no peace at all.'[3]

It has been argued that the 1929 Geneva Protocol, an international treaty, put poison gas into a special category of horror and therefore committed its signatories to avoid its use. This, it was believed, would be an effective ban to chemical weapons. Unfortunately, this turned out to be untrue, for an examination of warfare since the beginning of the treaty reveals chemical agent development and use up to the present day. What was even more alarming was the apparent escalation and proliferation of chemical warfare in the closing years of the twentieth century. The Chemical Warfare Convention was another mechanism, introduced in 1997, to try to prevent this proliferation. Nevertheless, it seems chemical agents could still play a significant role in future military conflicts because their tactical and strategic effectiveness outweigh all existing legal and moral restraints.

Certainly, the end of the First World War brought with it an increase in research and development of chemical warfare agents. Phosgene and mustard, developed and used by Germany and Britain, were to signal the start of this ultimate weapons race. In 1918, a team based at the Catholic University in Washington DC discovered Lewisite, a blister agent which was faster acting than mustard gas and caused immediate 'excruciating pain in the skin, sneezing, coughing, pain and tightness in the chest on inhalation, often accompanied by nausea and vomiting'.[4] These developments were followed by refinements of previously discovered agents. Hydrogen mustard, phosgene oxime, hydrogen cyanide, cyanogen chloride and others came out of secret laboratories such as Porton Down in Britain and Edgeworth Arsenal in the United States. In 1936, however, Dr Gerhard Schrader, a German scientist researching organic phosphorus compounds for a more effective insecticide, discovered tabun, the first nerve agent, which acted rapidly, was colourless, practically odourless and could poison the body by either inhalation or penetration of the skin. A new chapter in the history of chemical warfare had begun.

Historical accounts of military conflicts since the First World War lead us to believe that the use of chemical weapons was non-existent for legal or ethical reasons, or for fear of retaliation. Matthew Meselson, a noted biochemist, stated, 'There have been only two instances of verified poison gas warfare since 1925 ... in Ethiopia and Yemen.'[5] However, a preponderance of evidence exists to indicate that there have been numerous instances of chemical warfare use in military conflicts since 1918. In 1919, in India, stocks of phosgene and mustard gas were sent out from Britain for use on the frontier, and the Royal Air Force (RAF) is alleged to have used gas bombs against the Afghans in 1920. By 1925 the French and Spanish were employing poison gas in Morocco, and it had become clear that chemical warfare had found a new role, as a tool by which major powers could police rebellious territories.

In 1933 the Japanese established The Army Chemical Warfare School at Narashino, 21 miles east of Tokyo. The 11-month course ran for 12 years and turned out over 3000 chemical warfare officers for the Japanese Imperial Army. There is little doubt that from 1937 onwards the Japanese made extensive use of poison gas in their war against the Chinese.

Allegations included that the Japanese used mustard gas to drive Chinese peasants from caves and tunnels, and in 1938 China made a formal complaint against the Japanese to the League of Nations.

In 1935 the Italians invaded Abyssinia (Ethiopia) and over 700 tons of mustard gas was shipped for use by the Italian Air Force. Torpedo-shaped 500-lb bombs, with time-delay fuses, were also utilised. These bombs burst about 200 ft above the ground scattering spray over a considerable area. Later, aerial spraying was the preferred method. 'Groups of 9 to 15 aircraft followed one another so that the liquid issuing from them formed a continuous fog...soldiers, women, children, cattle, rivers, lakes and pastures were drenched continually with this deadly rain.'[6]

Conversely, the non-use of chemical warfare agents in the Second World War is looked upon by many as an example of the effectiveness of international legal and moral restrictions as exemplified in the Geneva Protocol. After all, this major world conflict offered many opportunities for poison gas use. Why then was it not utilised? There are, of course, a variety of reasons but certainly none of these were influenced by the legalities or ethics of the Geneva Protocol. It is clear, however, that the international reaction to the use of chemical warfare since the Second World War proved the existing legal and moral constraints had lost their effectiveness. Chemical weapons were a relatively cheap way to kill people, had a devastating effect on morale and accomplished their objectives without the destruction of buildings, equipment or land. These advantages, when weighed against the political consequences incurred by the use of chemical warfare agents, came out victorious every time.

After 1945, systematic chemical surveys continued, together with a search for novel agents based on advances in toxicology, biochemistry and pharmacology. The chemical industry, not surprisingly, was a major source of possible agents since most of the new chemical warfare agents previously developed had initially been identified in research on pesticides and pharmaceuticals. However, few candidate chemical warfare agents satisfied the special requirements of their potential users during this period. Of the many hundreds of thousands of chemicals screened between 1915 and 1953, only sixty were used in chemical warfare after 1915 or stockpiled for possible use as weapons in the future.[7] Two-thirds of them were used in the First World War when battlefields also served as testing grounds. However, fewer than twelve of the chemicals were found to be effective, and at any rate, these were supplemented or replaced by a similar number of more-developed chemicals after 1945.

Such chemicals can be divided into two categories: lethal chemicals, in other words, those designed to kill or injure the enemy, and those used to incapacitate the enemy. Before the Chemical Warfare Convention (CWC) was adopted in 1993, chemicals were selected as chemical warfare agents primarily because they had characteristics that made them so aggressive that munitions disseminating them would be competitive with conventional weapons. Lethal chemicals known to have been developed into chemical warfare agents can be divided into two further groups: tissue irritants and systematic poisons. The first group contains the choking gases (lung irritants or asphyxiates) and the blister gases (vesicants), and the second group contains the blood and nerve gases.

Chlorine, an asphyxiate, was the first lethal chemical to be used in the First World War. Widespread use of phosgene and diphosgene followed, and hydrogen cyanide was also produced. However, hydrogen cyanide was found to be unsuitable, as its physical properties (it was lighter than air) proved poorly suited to the munitions of relatively small payload capacity that were characteristic of most of the delivery systems of that time. Another significant development was that of agents such as mustard gas and the arsenical vesicants, for example Lewisite, which damaged the skin and poisoned through skin penetration.

Among the many new chemicals reviewed for their chemical warfare potential during the 1920s and 1930s were bis (trichloromethyl) oxalate, a congener of phosgene, and chloropicrin. Other chemicals examined included disulphur deca-fluoride, various arsenical vesicants, nitrogen mustards and higher sulphur mustards, cadminium, selenium and tellinium compounds, and carbonates. A few were found to offer some advantages over existing chemical warfare agents for particular purposes and were put into production. None, however, was thought superior to phosgene or mustard gas in general utility, and it was these two agents that formed the bulk of the chemical weapons stockpiled at the start of the Second World War.

The purpose of this book is to contribute to informed debate by providing an analysis of the development and deployment of chemical weapons from 700 BC to the present day. In Chapter 1 the groundwork for this, which follows a brief appraisal of historical prededents, is laid in a discussion of chemical warfare during the First World War, from which certain aspects are taken up and their development over subsequent years described. Chapter 2 examines the First World War in detail since it remains the most significant experience of the chemical threat. It contains some technical descriptions and a number of wider themes that have present-day relevance. One such theme is the nature of the whole development process for chemical weapons. Different stimuli have operated at different times. Chemical weapons came initially from chemists anxious to put their own particular expertise at the service of national war efforts. Later on, as the development process became institutionalised in different countries, the stimuli became more varied. They included, for example, known weaknesses in enemy protective equipment, the availability of new weapons delivery systems, the requirements of changing patterns of warfare, and the inevitable tendency towards self-preservation and propagation displayed by any institution. It seems clear that the enthusiasm of the chemists involved often outstripped that of the armed services. Nevertheless, the destructive potential of chemical weapons has compelled the military to pay attention to them, however much they dislike the notion of chemical warfare. Subsequent chapters analyse chemical warfare in the Second World War, the Cold War, the Korean War, the Vietnam War, the Middle East, Afghanistan, Bosnia, the Iraq–Iran conflict, the Gulf War and the 2003 invasion of Iraq. These chapters have two purposes. The first is to provide a catalogue of instances when the use of chemical weapons has been alleged. The second is to describe the military rationale underlying their use in those cases where the fact of their deployment is beyond reasonable doubt. Many of the chemical warfare allegations seem improbable and yet in no case cited in this book is there enough evidence to exclude them from a list of instances in which chemical warfare agents might have been employed. Certainly there have been five adequately substantiated instances of chemical warfare in the past 90 years: during the First World War, the Italian invasion of Ethiopia, the Japanese invasion of China, by the United States forces in Vietnam, and the Iran–Iraq War. It is also believed that large numbers of chemical weapons were deployed during the Yemeni Civil War and in the Korean War.

There are two points that emerge from these chapters that are worth drawing brief attention to here. First, it is clear that in those rare cases since the First World War when chemical weapons have been used on a substantial scale, it has always been against an enemy known to be deficient in anti-gas protective equipment or retaliatory capability. Second, in all substantiated cases of chemical warfare during the twentieth century, the employment of chemical irritants, such as tear gas, has always preceded the resort to more lethal chemical agents. This is true for the First World War, the Italian invasion of Ethiopia, the Japanese invasion of China and the Yemeni Civil War. In Vietnam, where irritants were used on a scale approaching that of the First World War, the reports of uses of more lethal chemicals remain unsubstantiated. These points seem to suggest that chemical weapons are likely to be militarily attractive only in strongly unequal conflicts, and that use of chemical irritants in war carries a risk of introducing more lethal forms of chemical warfare.

Chapter 3 is concerned with the period between the two world wars. It describes the ways in which public opinion in the field of chemical warfare was aroused after the experience of the First World War, and to some extent how public opinion was then exploited. The chapter considers some of the effects of this including how it stimulated the negotiation of the 1925 Geneva Protocol, one of the most important pieces of conventional international law prohibiting the use of chemical and biological weapons. The chapter also considers the national policies and programmes relating to chemical warfare in the inter-war period and examines important chemical warfare discoveries in these decades.

Chapter 4 deals with the Second World War. The non-use of chemical weapons was surprising, for by the end of the war the total stocks of chemical weapons by the belligerents far exceeded the total consumption during the First World War. The chapter explores the incentives there might have been for the different belligerents to use chemical weapons at different stages of the war, and then contrasts these with the constraints that might have been operating. The impressions that emerge are that the incentives to use chemical weapons seem to have been strongest in those cases where a belligerent's homeland was directly threatened and it was to their advantage to reduce enemy mobility, against blitzkrieg advances on land or against amphibious landings from the sea. But in these, and all other cases, the temptation was rejected. The reasons for this restraint varied from country to country, but included the fear of retaliation against other fronts and against civilian populations, personal opposition to chemical warfare on the part of political leaders and, in certain combat zones, the absence of trained soldiers and large supplies believed necessary to sustain a chemical warfare campaign.

Chapter 7 examines the threat (real and imagined) from a chemical warfare attack today by rationally assessing to what extent terrorist groups around the world are capable of making and using such weapons. Finally, throughout the book, the various protocols that attempted to bring about either the non-production or destruction of chemical weapons from 1675 to 1997 are examined and evaluated.

In his final report to Congress in the aftermath of the First World War, General John J. Pershing stated: 'Whether or not gas will be employed in future wars is a matter of conjecture, but the effect is so deadly to the unprepared that we can never afford to neglect the question.'[8] The First

World War generals were the last field commanders to actually confront chemical agents on the battlefield. Today, in the light of a significant terrorist chemical threat and solid evidence of the utilisation of chemical warfare in lesser conflicts, it is by no means certain they will retain that distinction.

Foreword

The expression 'the fog of war' is one that is very applicable to chemical warfare. There are huge uncertainties about the effectiveness of chemical weapons particularly in comparison with precision-guided high explosives. There seems little justification for describing chemical weapons as weapons of mass destruction for in no way do they threaten devastation on the scale of nuclear weapons.

There are some facts that are indisputable such as the use of toxic chemicals in the 1914–1918 World War and more recently in the Iran/Iraq conflict. Although many casualties resulted, the numbers were not exceptionally high in comparison with those from more conventional warfare. Also the use of chemicals was not decisive even in the World War against poorly protected soldiers in a trench warfare scenario in which chemical weapons might be expected to be most effective.

There are many other reports of the use of chemical warfare where the facts are sparse or where the allegations of use are not substantiated. Sometimes it is difficult to know whether the effects described result from poisoning or the fear of poisoning.

Today, the uncertainties about the effects of chemical weapons are increased by the large number of chemicals that are available – some are synthetic, some occur naturally, some cause irritation and incapacitation, some are instantly lethal, whereas others are active towards plants and animals. Also although the actual toxicity of chemicals is important (that is the smaller the dose to produce a required effect the better) other factors such as ease of synthesis, storage stability, ease of dissemination and persistence also play a major role. The uncertainty about the effects is further increased by the fact that against some chemicals soldiers have good defence from protective clothing, detectors and alarms, and medical countermeasures.

Dr Coleman in her book has provided an excellent historical perspective of chemical warfare. We see the desire of some to achieve a unique military advantage opposed by others particularly revolted by warfare with poisonous gases. We see the reluctance of some military commanders to use weapons they did not fully understand. Was it the fear of retaliation in kind that prevented chemical warfare in the Second World War or was it the uncertainty of effects both on the user as well as on those attacked that was the most restraining influence?

Some may argue that it was the existence of the Geneva Convention that prevented chemical weapon use in the Second Word War but the size of the stockpiles in various countries does not really support this argument. When one looks at the size of the US and USSR stockpiles declared under the Chemical Weapons Convention 1997, it becomes apparent that very large quantities were considered necessary to be militarily effective. There is such a significant logistic burden in deploying such stocks that any commander would like to be confident about the outcome. This may be another reason why chemicals were not used.

Dr Coleman comments on the use of chemicals by terrorist groups. The wide range of options are described as well as the difficulties. There is little doubt that the potential use of poisonous chemicals creates much fear and apprehension. It is important however not to exaggerate the threat. Although as is often stated in popular reporting that a few grams of some chemicals can kill some thousands of people, the problem of bringing those thousands into contact with the few grams are so great that it is unlikely that any terrorist chemical incident would have more than quite local effects. We must ask then, whether, bearing in mind the public revulsion to poisonous chemicals, few terrorist groups would wish to deviate from more conventional methods. Nevertheless, it is important that both national and local governments take steps to have well-trained people to provide proportionate response in the event of a chemical terrorist attack. The advice that should be available to the public at large must be based on well-founded data and should not alarm the public unnecessarily.

Dr Coleman provides some apt comments about the Chemical Weapons Convention. In many respects this is a unique arms control treaty since it attempts to ban the production and stockpiling and the use of all chemicals except for permitted purposes. In other words it impinges on the activities not just of defence organisations but also of worldwide chemical, pharmaceutical and biotechnology industries. Signatories to the convention were required to agree to the destruction of any stocks of chemical weapons and to report on and allow inspections of all sites in their countries where chemicals listed in the schedules were made. It is perhaps astonishing that the whole process that is overseen by the Organisation for the Prohibition of Chemical Weapons works as well as it does. And yet it is difficult to believe that an international treaty such as this, even if each and every country had signed (and the small number of non-signatories are significant), could be monitored inter-nationally to prevent any state party or any terrorist group from covertly breaching the convention. Few countries fully appreciate that signing the convention requires national authorities to take steps to ensure all the provisions of the convention are adhered to. Too few politicians and probably too few members of the community that work with chemicals on a daily basis properly understand the far-reaching requirements of the convention and the national responsibilities under it. Concerns about chemical terrorism are raising the profile of the issues but more is required.

The history of chemical warfare is important so we can learn lessons for the future. Dr Coleman has provided a well-referenced account of the history. Readers should form their own judgements on the threat chemical weapons pose – whether they really are weapons of mass destruction, their attractiveness to terrorists and the strengths and weaknesses of the Chemical Weapons Convention.

Dr Thomas D. Inch OBE; BSc, PhD, DSc, FRSC

Formerly, until retirement in 2000, Secretary-General and Chief Executive, Royal Society of Chemistry.

Previously, until 1993, Vice President responsible for R&D for BP in North America.

Before 1985, spent 20 years at Porton Down in research with over 100 publications on chemistry and medicinal chemistry. During the latter part of this period, occasionally served as a technical advisor to the UK government in discussions leading to the Chemical Weapons Convention.

Became Chairman of the advisory committee to the UK National Authority to the CWC following the passing of the UK Chemical Weapons Act.

1
Historical Precedents?

War is defined as a state of hostility, conflict, antagonism or struggle between two opposing forces for a particular end. When chemical weapons are added to an existing arsenal, the nature of the conflict is changed in two significant ways. First, the number of deaths and injuries are potentially increased. Secondly, if one country has chemical weapons this causes other countries to devote vast resources to develop a matching arsenal. Since the invention of these weapons, warfare and the threat of warfare has never been the same. The twentieth century saw the development of progressively more deadly chemical weapons. It saw their use, with significant effect, in a world war, in a regional conflict and the first instance of their use by terrorists.

Today, the so-called weapons of mass destruction (WMD), which is an umbrella term to include chemical, biological and nuclear weapons, have become one of the most prominent topics in the news since the events of September 11, 2001. Not a day passes without much being said about them in the media, by politicians and other commentators. The world's leaders continually warn us of the dangers of WMD. The Secretary-General of the United Nations, Kofi Annan, called the possible terrorist use of chemical, nuclear and biological weapons, 'The gravest threat the world faces.'[1] Western leaders, especially George W. Bush and Tony Blair, have told us that international terrorists and the states that support them are today's greatest threats to national and global security. War, they argue, is necessary and justified to remove these threats because unless the regimes in the accused countries are changed, WMD may be used with devastating effects. Should we believe these prophecies of doom or are they exaggerated nightmares? It is impossible to judge the threat unless we know the answers to some key questions. What are WMD? How do chemical, biological and nuclear weapons differ from each other? What are the effects of the use of these weapons? Which terrorist groups are capable of making and using these weapons? What facilities do countries need to manufacture and deliver WMD?

Weapons of mass destruction take chemical, biological and nuclear form. The use of the term is recent; it is also controversial. The Royal United Services Institute, for example, point out that North Atlantic Treaty Organisation (NATO) still uses the 'nuclear, biological and chemical' description as each type of weapon is distinctive. In effect, it seems that the creation of the blanket acronym WMD blurs these distinctions. Simply put, nuclear, biological and chemical weapons are designed to kill and injure a large number of people: Nuclear weapons have the purpose of destroying much of the enemy's property, particularly its cities and industries; biological weapons spread disease deliberately in human populations; and chemical weapons are designed for the effective dispersal of a chemical warfare agent, for example gas in 1914 and sarin, or potentially VX, today. This book is concerned with the history of the development and deployment of chemical weapons and, as such, the focus hereafter lies in that area.

Most of the weapons used today are chemical. The explosion of tri nitro toluene (TNT) is a chemical reaction and so is the combustion of the nuclear bomb. This book, however, is concerned with those weapons that are based on the toxic properties of chemicals rather than on the energetics of their interaction. As a category, toxins have recently acquired greater prominence in the literature on chemical and biological warfare, though not because of any increase in their potential for weaponisation, despite their being among the most dangerous substances known today. It is true, however, that some toxins are becoming more accessible to quantity production than they once were. 'Toxin' is a word that has no commonly accepted meaning in scientific literature. The 1972 Toxic Weapons Convention (TWC) covers 'toxins whatever their origin or method of production'. The TWC does not define toxins, but its *travaux préparatoires* show that the term is intended to mean toxic chemicals produced by living organisms. Toxins, of course, are both toxic and chemical in nature. According to the Chemical Warfare Convention (CWC), toxic chemical refers to any chemical that through its action on life processes can cause death, temporary incapacitation or permanent harm to humans or animals. Some toxins, although toxins are usually associated with biological warfare, are included in Annex 1 of the CWC. So, although there is no consensus on the term 'toxin' among scientists, international law regards a wide range of substances as toxins. Indeed, Schedule 1 of the CWC lists 'ricin', a toxic glycoprotein derived from the castor oil plant, in its forbidden substances. A weapon system based on toxic chemicals may be looked at as the sum of four parts: a system to deliver the munitions; munitions to disseminate the chemical agent; the agent itself; and the part played by the environment in transporting the disseminated chemical to its target. Each of the four parts is dependent to a greater or lesser extent on the other parts. For example, if the attacker is relying on the atmosphere to transport the agent to the target's lungs, the agent chosen must be one which can be made airborne in a form which will penetrate the lungs. If the chosen agent is one that is sensitive to heat, then the chosen munition must avoid it.

Chemical warfare means the wartime use, against an enemy, of agents having a direct (toxic) effect on man, animals or plants. The use of chemical warfare agents against man, rather than animals or plants, is referred to as gas warfare, even though the substances used may be solid, liquid or gaseous. The toxic effects produced in gas warfare may be transient or permanent, ranging from a temporary irritation of the eyes to death.[2] There are four main categories of chemical warfare agents: choking, blister, blood and nerve agents. Choking agents, such as carbonyl chloride or phosgene, attack the respiratory tract making the membranes swell and the lungs fill with fluid so that the victim drowns in his own juices. Choking gases are the classical agents of chemical warfare but are unlikely to be used in a modern chemical war as their initial irritancy or smell immediately warns of their presence, and gas masks can therefore be put on before a lethal exposure. In addition, the toxicity is nowadays too low; for example, the lethal exposure to phosgene is around $3200 \, mg/min/m^3$.[3] The best-known blister agent is mustard gas, also called 'Yperite'. Mustard gas is a persistent agent that remains toxic for a long period and can be lethal. Blister agents produce large watery blisters on exposed skin that heal slowly and may become infected. Blister agents may also damage the eyes, blood cells and respiratory tract. There are two main classes of blister agent – arsenicals, such as Lewisite which has a sharp, irritating odour and causes immediate eye pain,[4] and mustards. Arsenicals give enough warning of their presence for protective clothing to be put on in time, mustards do not, and it is for this reason they are still in arsenals today. Agent Q, one of the most lethal variants today was in fact discovered in the 1960s, and will blind at an exposure of less than $50 \, mg/min/m^3$ and kill if inhaled at dosages of $200 \, mg/min/m^3$. Indeed, Agent Q's inhalation toxicity approaches that of the nerve gases. Blood agents such as AC are absorbed into the body by breathing and kill by entering the bloodstream and causing vital organs to cease functioning. There are two main groups of nerve agents, the G-agents, typically volatile liquids that break down quickly and cause death when inhaled, and the V-agents which are much more persistent and can be absorbed through the skin. The most lethal nerve agents are three G-agents, tabun, sarin and soman, and a V-agent, VX. Tabun was first discovered in 1936. It is a colourless liquid with a fruity smell, first produced in industrial quantities in Silesia in 1942. Sarin was also discovered in Germany in 1938. It is a colourless liquid with no smell. Soman, again discovered in Germany in 1944, is also a colourless liquid with a fruity smell. Tabun is about half as toxic as sarin, and soman, about twice as toxic.[5] It is a moot point whether tabun is still considered worth stockpiling as its toxicity is not as high as the other G-agents, but it has a persistency in the field which may be considered tactically useful in that it could provide a vapour hazard for some days after dissemination. A subsidiary of Imperial Chemical Industries (ICI) in Britain and Bayer in Germany, both working independently, discovered VX in the early 1950s.[6] It too is a colourless liquid with no smell. Nerve agents are organophosphorus compounds (as are, for example, insecticides). In the body they prevent acetyl-cholinesterase, an enzyme essential for the normal functioning of the nervous system, from acting normally. The initial symptoms vary according to which agent is absorbed. A low dose of any nerve agent will generally cause reactions like a running nose, contraction of the pupils, blurred vision, slurred speech, nausea and hallucinations. A high dose will cause the victim breathing problems, convulsions, deep coma and finally death. At even higher doses, the symptoms will occur very rapidly and the person will die from suffocation as both the nervous and the respiratory systems fail at the same time. A minute drop of a nerve gas, inhaled or absorbed through the skin or eyes, is enough to kill within about twenty minutes.

The job of a chemical munition is to create a toxic environment over as much of the target as is compatible with the toxicity of its charge. It must convert its bulk load either into an even distribution of liquid or solid particles, or into a cloud of vapour, or into both. It must, additionally, do this in a certain time. These are strict demands, and they are made more severe by the diversity of chemical agents now in stockpiles. Each agent has a combination of physical characteristics and toxic behaviour that is unique but, nevertheless, all munitions work on the same basic principle: they cause the transfer of energy from a store, generally an explosive, to the chemical load. The simplest chemicals to disperse are the volatile, non-persistent ones such as phosgene; the hardest ones are the heat-sensitive solids that include such things as ricin, a protein more toxic than nerve gases and about which much has been said in recent years.

One of the assets of chemical warfare is that it does not depend on extraordinary delivery systems. Chemical munitions may be adapted for delivery by almost any means – grenade throwers, artillery and aircraft. Indeed, in some cases the delivery system may be the environment itself – the chlorine cylinders of the First World War for instance. However, once a chemical weapon has been deployed its user has no further control over it. This, of course, is true for any other weapon, but whereas the effects of, say, high explosive follow within a fraction of a second of detonation, those of a chemical may be delayed for minutes, hours or even days. In this lies both the strength and the weakness of chemical warfare. On the one hand, a toxic atmosphere may be set up which will envelope the whole target area, seeping into tunnels, bunkers and buildings. On the other hand, the entire loads may be blown uselessly away by a sudden wind. Certainly, the weather, winds and, to some extent, precipitation and indeed the practical limitations of dispersal generally limit the use of chemical weapons against concentrated targets as opposed to large geographical areas. Chemical weapons can be very effective against troop concentrations, military facilities and highly populated civilian areas. However, chemical weapons do not, obviously, pose much of a threat to a geographically dispersed civilian population. It must be emphasised then that no matter how well-designed a chemical weapon is, its effectiveness depends critically on the prevailing weather conditions. All this implies that previous knowledge of target conditions is essential to a chemical attack. It appears, therefore, that it was not for nothing that the 12th Earl of Dundonald consulted the Meteorological Office in 1914 before revealing to Lord Kitchener his grandfather's plans for chemical warfare.[7]

Chemical weapons are not a new method of warfare, they have been in recorded use since about 2000 BC. However, science and technology have refined these weapons and now their potential is awesome. It was the rise of the modern chemical industry at the end of the nineteenth century that first made feasible the use of significant quantities of toxic chemicals on large-scale battlefields and, indeed, chemical weapons were first used on a significant scale by both sides in the First World War. They were then used immediately after the war by Britain in Iraq (1920), and Spain in Morocco (1921). They were also used by Italy during its invasion of Abyssinia (Ethiopia) in 1935–1936, Japan during its war against China in 1937–1943, and by the United States in Vietnam in 1965–1975. Both sides in the Iran–Iraq War used them in 1980–1988, and in a particularly high-profile attack they were deployed by Iraq against the Kurds at Halabja in 1988.

However, the use of poisons that could be considered chemical weapons dates back to antiquity. The wars of ancient India in about 2000 BC were fought with smoke screens, incendiary devices and toxic fumes that caused sleep. Thucydides tells of the use of gas during the Peloponnesian War (431–404 BC); also the use of an incapacitating agent, one which caused incessant diarrhoea, is recorded by Polyaenus, Fronto and Pausanias. The Spartans used arsenic smoke, comprised of pitch and sulphur, during the sieges of Plataea and Delium. The pitch and sulphur were ignited and the consequence was 'a fire greater than anyone had ever yet seen produced by human agency', the Greek historian wrote.[8] There is some debate concerning the effect of this new weapon on the final outcome, but it is unequivocally true that even the crudest chemical weapon will create fear and panic. Undoubtedly, this was exactly what happened during both sieges, making the way then clear for the Spartan Army to seize the advantage presented to them by the incapacity of their enemy, an opportunity they did not squander. Between 82–72 BC the Romans used 'toxic smoke' against the Charakitanes in Spain, causing pulmonary problems and blindness not dissimilar to the effects of phosgene centuries later. In this case the effects of this chemical weapon are clear – the Charakitanes were defeated in two days.

Almost a millennium later at the siege of Constantinople (AD 637), the Byzantine Greeks employed 'Greek Fire', a weapon invented by an architect, Callinus of Helipolis, which became decisive at this time and was used with success by the Byzantines in their campaigns up to the thirteenth century. Indeed, it can be argued, its effectiveness was a prime reason for the long survival of the Byzantine Empire. The exact composition of Greek Fire is still a mystery but naphtha or petroleum is thought to have been the principle ingredient, probably with sulphur or pitch and other materials added. Indeed, Greek Fire, it can be assumed, was the forerunner of Napalm. It is not clear, however, how it was ignited, but quicklime was probably used, mixed with the main ingredients at the last moment. Once lit, the substance was very hard to extinguish; water was useless, sand or vinegar was the only solution.

In the Middle Ages, chemical warfare was put to similar use as at the siege of Delium and such usage continued through to the fifteenth century. In 1456 an alchemist who prepared a poisonous mixture saved Christian Belgrade from the attacking Turks. The Christians dipped rags in the chemical and burned them, creating a toxic cloud that was not dissimilar to the chlorine clouds on the Western Front in 1915. This drifting cloud attack with an arsenical smoke is described by the Austrian writer, von Senfftenberg, with the comment: 'It was a sad business. Christians must never use so murderous a weapon against other Christians. Still, it is quite in place against Turks and other miscreants.'[9]

The 'Notebooks' of Leonardo da Vinci reveal a design for a chemical weapon which comprised a mixture of powdered arsenic and powdered sulphur packed into shells and fired against ships. Such a weapon was indeed developed and deployed, and as such is the first recorded usage of a chemical weapon.[10] This use provided a precedent for the use of poison bullets against enemies and also led to the first attempt to prohibit the use of chemical weapons. This was elaborated in the Strasbourg Agreement (27 August 1675), a bilateral French and German accord which directed that neither side should use poison bullets and, as such, constitutes the first international agreement in modern history in which use of such weapons was prohibited.

As chemistry advanced during the nineteenth century, many new proposals for chemical weapons were made; for example, organoarsenical bombs and shells at the time of the Crimean War and a chlorine shell and other devices during the American Civil War. Indeed, Napoleon III is said to have put hydrogen cyanide to military use in 1865.[11] An influential figure in the nineteenth-century history of chemical warfare was Thomas Cochrane. In March 1812 Britain's prince regent, the future George IV, received from Cochrane a proposal aimed at undermining the power of Napoleon in a manner guaranteed to revolutionise the rigid customs of warfare. At that time the Duke of Wellington was struggling through Spain and the strength of the Royal Navy was being sapped by the need to maintain a tedious blockade of the key ports where Napoleon's warships waited for an opportunity to escape into the Atlantic. Cochrane's proposals, which the prince turned over to his advisors, offered a radical scheme by which a beachhead on the coast of France could be gained quickly and decisively. Cochrane detailed two new innovative weapons systems, the 'explosion ship' and the 'sulphur ship' or 'stink vessel'.[12] The plan stipulated that the two weapons were to be used in conjunction with each other. First, the explosion ship would be towed into place at an appropriate distance from anchored enemy ships, heeled to a correct angle and anchored. When detonated the immense explosion would cause debris to fall onto the enemy causing mayhem. Then the follow-up, the sulphur ship would be towed into place and when the wind blew windward charcoal covered with sulphur would be ignited. The resulting clouds of 'noxious effluvia', as Cochrane termed them,[13] were expected to be pungent enough to reduce all opposition as the defenders ran away to escape the choking gas. A quick landing by the British could then secure an otherwise unattainable position and clear the way for the establishment of a beachhead. Thomas Cochrane had prefaced his plan thus: 'To the Imperial mind, one sentence will suffice. All fortifications, especially marine fortifications, can undercover of dense smoke be irresistibly subdued by fumes of sulphur kindled in masses to windward of their ramparts.'[14] He had, in fact, been partly anticipated by a good two millennia. The Peloponnesians had attempted to reduce the town of Platea with sulphur fumes in the fifth century BC. At length, an expert panel decided there was merit in this unusual scheme, but fear of the implications that such radical devices would have on warfare stifled their enthusiasm. What would happen, they asked, if the enemy gained knowledge of this new technology and turned it against Britain's defences?[15] The proposal was rejected on the grounds, 'It would not accord with the feelings and principles of civilised warfare.'[16]

Nearly 40 years later, in July 1853, Cochrane, now 79 years old, urged the First Lord of the Admiralty, Sir James Graham, to reconsider the King's 1812 decision and use the explosion and sulphur ships at Sevastapol as the possibility of war in the Crimea increased. Again, the idea was quickly dismissed. A year later, in July 1854, Cochrane again urged Graham to employ his vessels to force the Russian troops away from the fortifications of the harbour at Krondstadt. He said that once the ships had exploded and the enemy was scattered a British landing could be made and the enemy's guns, once captured, could be manned and turned on the Russian ships anchored below the batteries. Once more, however, the scheme was rejected and the British sailed to the Baltic where they eventually failed to subdue Krondstadt.

Throughout the debate, the details of the scheme remained secret. In the boardroom at the Admiralty, the plan showed the sulphur ships with layers of coke and sulphur ready to emit their choking fog. Added to the scheme was the intention to create a smoke screen by pouring naphtha onto the surface of the harbour and igniting it with potassium,[17] perhaps a nineteenth-century version of Greek Fire. Cochrane was convinced that a few hours would accomplish what months of debilitating conventional warfare had failed to achieve. Palmerstone's government appeared to be close to sanctioning the strategy when Sevastopol was taken in September 1855, followed soon by the end of the war. All discussion of the revolutionary weapons was dropped, and the plans were sealed away on the shelves reserved for confidential matters at Whitehall. Cochrane died in 1860 and his secret war plan remained secure until 1908 when Palmerstone's correspondence was published. Less than a decade later the sulphuric yellow clouds of mustard gas ravaged thousands in the trenches of France.

A few years after Cochrane's death, as the American Civil War drew to an end, Ulysses Grant's army was stalled outside Richmond during the siege of Petersburg, Virginia (1865). A plan was devised to attack Confederate trenches with a cloud of hydrochloric and sulphuric acids.[18] This plan was not acted upon but this idea, along with Cochrane's proposals, proved to be a prerequisite for the Declaration of St Petersburg (1868). This declaration renounced the use of explosive projectiles charged with fulminating or inflammable substances in war. Additionally, it prohibited 'material of a nature to cause unnecessary suffering'.[19] Twenty signatories participated of which Britain, France and Germany are still adherents.

By the end of the nineteenth century the use of poison gas was still by far the exception and not the rule in war, and yet there were in all the great powers a number of men who foresaw its widespread use should a general conflagration engulf Europe.[20] Indeed, a concern with poison gas manifested itself at the Hague Conference of 1899. One of the agenda items dealt with prohibiting the use of shells filled with asphyxiating gas: 'The contracting Powers agree to abstain from the use of projectiles the sole object of which is the diffusion of asphyxiating or deleterious gases.'[21] The proposed ban eventually passed with only one dissenting voice, that of the American representative, naval Captain Alfred T. Mahan, who declared that,

> It is illogical and not demonstrably humane to be tender about asphyxiating men with gas, when all . . . admit it is allowable to blow the bottom out of an ironclad, throwing four or five hundred men into the sea, to be choked by water, with scarcely the remotest chance of escape.[22]

For Mahan, it made no sense for the United States to deprive itself of the ability to use, at some later date, a weapon that might prove to be more humane and effective than anything then present in the American arsenal.

The Hague Conference did not prevent some nations from discussing the use of chemical weapons, and at least one country, France, experimented publicly with gas. The French Army tested a grenade filled with ethyl bromoacetate, a non-toxic tear agent (or lachrymatory) developed for use in the suppression of small-arms fire from the concrete casements then prevalent in the fortifications that dotted western Europe. In 1912, French police used 26-mm grenades filled with this agent to capture a gang of notorious bank robbers. The British and the Germans, unlike the French, did not experiment with chemical agents for military use, but, nevertheless, at the outbreak of the First World War Germany's highly advanced dye industry gave it a technological base from which it was able to easily develop weapons of this nature.

2
The First World War

Gas! Gas! Quick, boys! – An ecstasy of fumbling,
    Fitting the clumsy helmets just in time;
But someone still was yelling out and stumbling.
    And flound'ring like a man in fire or lime...
Dim, through the misty panes and thick green light.
    As under a green sea, I saw him drowning.
    In all my dreams, before my helpless sight,
He plunges at me, guttering, choking, drowning.

        – Wilfred Owen, *Dulce et Decorum Est*[1]

The most persistent assumption underlying the decisions taken by the great powers in July and August 1914 was the illusion that the ensuing war would be short. The thinking behind this was relatively simple: modern methods of transportation and communications created unprecedented opportunities for speed and mobility in attack. In fact, all the war plans of the great powers before 1914 hinged on railway timetables and the rapid deployment of men in the field. Indeed Kaiser Wilhelm II assured his troops they would be 'home before the leaves fall' and certainly troops of all nations believed 'it will all be over by Christmas'. Young men went off adventurously, glad to change their lives, to travel. They were answering the call of duty and were sure they would soon be back home crowned with victory; in London, Berlin and Paris they left singing and exuberant. But, the dream became a nightmare. The belief in speed was crucial. The most famous stratagem, the German Schlieffen Plan called for a lightning attack on France – but this was not exceptional. France had Plan 17 which proposed a quick strike through Alsace; Russia's Plan B called for Russia to seize the offensive and attack through Poland and Britain's planning for the

British Expeditionary Force (BEF) assumed it must land in France within days of war being declared for it to be effective. These plans show the extent to which strategists committed themselves to the view that standing on the defensive would lead to destruction. These strategic calculations, however, proved to be ill-founded and the end of 1914 locked the armies on the Western Front in a deadly, static form of trench warfare and about to experience the nightmare intensification of the industrialised battlefield.

Unwilling to accept the deadlock of trench warfare, army staffs of both sides deliberated on ways to break the stalemate and return to open or manoeuvre warfare. Alternatives were proposed; some were strategic like the Allied attack on Gallipoli, some tactical like the change from full-scale bombardment prior to attack at Neuve Chapelle. In April 1915 the Allies carried out a military landing on the Gallipoli Peninsular. They were to hold the area and advance.[2] The Gallipoli Campaign is remembered as one of the classic failures of military history. Undoubtedly, however, the Campaign was of lasting significance because the troops, in the first action of its kind, under immense physical and emotional pressure, strove for political and military gains to no avail. At Neuve Chapelle British military doctrine dictated that indirect fire could cut wire and that a short bombardment would allow a break-in. However, indirect fire was inflexible and the artillery found it difficult to locate targets and so therefore, indirect fire was ineffective.[3] However, although the break-in was easily achieved, the breakthrough was defeated as the Germans had time to bring up reserves during the gaps between the phases of the battle. Therefore, both plans failed for a variety of reasons, and the deadlock on the Western Front continued.

By the autumn of 1914, interest in the combat possibilities of toxic chemicals had quickened. In the United Kingdom the Admiralty was reconsidering the proposal of Admiral Cochrane for the offensive use of sulphur dioxide clouds. In the United States a patent application was being prepared that related to an artillery shell charged with hydrogen cyanide. In France, army officers were considering the tactical possibilities of the tear gas weapons that the Paris police force had been using since 1911. In Germany a team was experimenting with phosgene and arsenical grenade fillings.[4] Yet, whatever the military authorities may have felt about this and other chemical warfare work, it is clear that during the early months of the fighting neither the war nor the technology had developed to a point at which the work could be usefully exploited. Toxic chemicals had no obvious part to play in the sort of fighting which took place during the opening campaigns and, in principle, their use was obnoxious to the professional code of the military, a distaste symbolised in the somewhat vague proscriptions that had emerged from the Hague Conferences.

As the German offensive became bogged down in trench warfare, the German military could no longer be certain of victory and it was in this atmosphere that the world's first chemical warfare assault was launched. If seeking developmental pre-conditions for the introduction of modern chemical warfare, there is no doubt they existed in Germany prior to the outbreak of the First World War. The industrial and indeed, economic markets of pre-war Germany were dominated by the chemical industry which had a tradition of intense and well-developed research and development programmes.

Considered uncivilised prior to the First World War, it could be argued that the development and use of chemical warfare was necessitated by the requirements of wartime armies to find new ways of overcoming the stalemate of unexpected trench warfare.[5] Old gases such as chlorine were used, and newer gases such as mustard were developed, and successfully used, as a terror weapon meant to instil confusion and panic amongst the enemy prior to an offensive.

On relatively few occasions in military history has an army employed weapons that are so intrinsically unreliable that they pose the same threat to the side using them as to the enemy. In such cases friendly casualties are almost inevitable and the best example of this process is the use of poison gas during the First World War. Subject to the vagaries of wind and weather, poison gas always posed a threat not only of blowing back onto the advancing formations of troops but also of gathering in thick clouds around the enemy trenches so that even if an initial attack was successful it was impossible for the attacking forces to occupy enemy lines without falling victim to their own gas.[6] Indeed, in the seventeenth century Siemienowitz warned: 'One must take care lest one suffers one-self from the means intended to injure others.'[7] He went on to remark that toxic projectiles 'do not give the effect expected of them since the poisonous cloud goes straight up into the air, and is dissipated by the wind'. He also perceived that foggy or rainy weather favoured chemical warfare.[8]

The beginning of the development of chemical weapons during the First World War was haphazard. The impetus came from chemists who had become aware of the noxious effects of certain chemicals in their laboratories, and who felt that these efforts could be exploited to assist national war efforts. Certainly from 1914 onwards, attempts were being made in several academic laboratories throughout Europe to convert laboratory chemicals into weapons of war. Nevertheless, it took some time for these initial efforts, in which a number of scientists succeeded in killing themselves, to produce significant results on the battlefield. Several of the belligerents had been using munitions filled with irritants from almost the beginning of hostilities and, although it is popularly believed that the German Army was the first to use gas it was in fact the French who initially deployed it. As early as August 1914 the French, experimentally, fired tear-gas grenades (xylyl bromide) against the Germans. This early venture into chemical warfare had little impact other than to draw attention to its potential. However, it was the Germans who were the first to give serious study to the development of chemical weapons and, ultimately, the first to use poison gas on a large scale. Despite the great psychological activity of obvious chemical warfare agents, the weapons designers of 1914 soon realised that it was no easy matter to design a weapon that could deliver effective dosages of the agent to an enemy deployed over a distant target area. It seemed the only practicable way of delivering an agent was to contaminate the enemy's surroundings, particularly the air he breathed, in the hope that some of the chemical agent would eventually penetrate his body. It was realised then that the performance of the potential weapon was crucially dependent on the state of the atmosphere. On the one hand, a great load of poison gas might be carried by the wind and permeate the entire target area, on the other hand, the whole load might be uselessly blown away or become so diluted as to be harmless. Certainly, it was recognised that the greater the dependence of any weapon system on the prevailing weather conditions, the fewer would be the occasions on which it could be used.

During the capture of Neuve Chapelle in October 1914, the German Army fired shells at the French that contained a chemical irritant whose result was to induce a violent fit of sneezing. This gas potential attracted the attention of the German High Command and consequently they asked the Kaiser Wilhelm Institute in Berlin to investigate the possibility of using a more effective chemical agent. The only guideline provided by the military was that the Hague Declaration (1899), banning project-iles used exclusively for delivering poison gas, had to be circumvented. Adhering to the letter, if not the spirit of the ban, the Germans devised a gas shell (T-Stoff) that also contained an explosive charge for producing a shrapnel effect.[9] Three months later, on 31 January 1915, the Germans employed T-Stoff shells for the first time, on the Eastern Front.[10] These were fired in liquid form contained in 5-cm Howitzer shells against the Russians at Bolimov. German officers who were confident that their new weapon would neutralise the enemy positions were therefore surprised when their attack was repulsed with severe casualties. The new experiment had proved unsuccessful as the tear-gas liquid had failed to vaporise in the freezing temperatures prevalent at Bolimov. Not giving up, the Germans tried again with an improvised tear-gas concoction at Nieuport against the French in March 1915, however with limited success.

The Germans quickly realised that the value of irritants increased with the scale on which they were used. If irritant agent harassment of enemy troop units was to disrupt supply lines or lower battlefield performance significantly, the Germans realised the agents would have to be used over a wide area and for prolonged periods. The scattering of a few irritant shells over enemy positions had only nuisance value, given the inefficiency of early weapon designs. Once the German High Command had become accustomed to thinking about and using irritants on a large scale, it was only a matter of time before it began to do so for more lethal chemicals as well. After the Battle of the Marne, the mobility of both armies had been destroyed by the appearance of trench warfare. With its armies dug in from Switzerland to the Channel ports, Germany had almost exhausted its pre-war stockpile of high explosives and to very little effect. Furthermore, the blockade at sea was depriving the country of the raw materials needed to manufacture explosives, primarily nitrates from Chile. At this point the German High Command became particularly ready to listen to the country's industrial chemists believing only they could resolve the ammunition crisis. Ludendorff, Chief of the General Staff at this time, told of a meeting attended by the heads of Krupps and the forerunners of IG Farben (the great German combine of the chemical industry which included Hoechst, Bayer and Badische Anilin- & Soda-Fabrik (BASF)), that held a virtual world monopoly not only in dyestuffs, but also in the majority of organic chemicals. The purpose of this conference was to reorganise munitions production and during it the representative of IG Farben promoted the idea of using chemical agents to injure or kill, rather than harass.[11] Gas was seen not as a substitute for explosives, but as a possible way of breaking through the stabilised front: an entrenched enemy was comparatively safe from projectiles but vulnerable to airborne poisons. The decision was accordingly taken to try chemical agents on the battlefield.

To find a more effective means of employing gas on the battlefield, the German High Command turned to Professor Fritz Haber, the world-famous chemist who had developed a crucial process for extracting nitrates from the atmosphere (Nitrogen Fixation). This process was used to manufacture fertilisers and later, once war broke out, explosives.[12] From 1915 Haber directed all the work on poison gas. Following the war his work was appreciated in the most unlikely quarters:

> It would be difficult to exaggerate Germany's debt to Haber, yet he never attained a higher rank than Captain. His country, however, made proper use of him. Had he been born an Englishman, he would certainly have attained higher rank, but almost certainly he would not have been properly utilised until it was nearly too late.[13]

Believing that T-Stoff shells did not provide a high-enough concentration of chemicals to produce enemy casualties, he suggested the use of large commercial gas cylinders as a delivery system. Cylinders could deliver large amounts of gas and, like the T-Stoff shell, did not technically violate the Hague ban on projectiles. Haber also recommended the use of chlorine as an agent because it was commercially produced and readily available in large quantities; additionally, it also satisfied the requirements for military application – it was lethal, effective, non-persistent and volatile.[14] Toxicologically, it was a powerful lung irritant producing death by asphyxiation.

The German High Command chose, from a study of prevailing winds, the most suitable part of the front for the experiment, and it was decided that the Ypres sector of the Western Front was to be the proving ground. Although the contrary has been argued, it seems doubtful whether the German High Command regarded the forthcoming chlorine attack as anything more than a battlefield trial of an experimental weapon. The local field commanders were not enthusiastic about gas, and their requests for augmentation of ammunition supplies and reserves to exploit such success as it might achieve were turned down.[15] Furthermore, it seems doubtful whether the German High Command expected startling results from the experiment, for it was apparently prepared to risk premature disclosure of the new weapon although justification of its proponent's claims depended on massive surprise. A belief in the superiority of the German chemical industry and the inability of its British and French counterparts to provide the means of retaliation would surely not have been adequate reason to take this risk. After all, chlorine was one of the simplest industrial chemicals to make, and indeed was being made in Allied factories, albeit only on a small scale in liquefied form. On 10 March 1915 German Pioneer Regiment 35 (referred to as the *Stinkpionere* by other German troops) had emplaced 1600 large and 4130 small cylinders containing 168 tons of chlorine in the area earmarked for the attack around the Ypres salient.[16] Then, for one month, the Pioneer troops sat and waited for the wind to shift westerly towards the enemy trenches in the Salient because only then could they safely unleash the chemicals by opening the cylinder valves.

With the coming of April the weather improved and late in the afternoon of 22 April 1915, with temperatures into the seventies, as a setting sun cast long shadows over the battle-scarred terrain around the Belgian City of Ypres, at 1724, three flares rose from an observation balloon over the German lines and burst against the darkening eastern sky, and German artillery commenced a new and furious bombardment of the towns and villages. The code word *Gott strafe Engelland* was passed along the line and German assault troops moved into position. Finally, signal 8888 – open the gas containers – was issued and the men of Pioneer Regiment 35 pulled on their masks, bent over their cylinders and wrenched open the cocks to release the gas into the wind. French sentries suddenly noticed two curious greenish-yellow clouds drifting slowly out across no-man's-land towards their line. These clouds spread laterally, joined up and, moving before a light wind, became a bluish-white mist. Rapid fire from the French 75-mm field batteries continued, as did the rifle fire of the Germans who appeared to be advancing. Puzzled, but suspicious, the French troops suspected that the cloud masked an advance by German infantry and ordered their men to 'stand to' – that is, to mount the trench fire step in readiness for probable attack. The cloud did not mask an infantry attack however, at least, not yet. The Canadian Division on the right of the French had only just arrived in the sector and as such had no proper communication with them. Additionally, all telephone lines to and from Divisional Headquarters had been cut by the German bombardment. Therefore, it was impossible for anyone to form a clear picture of what was happening but, as the minutes passed, people in the rear areas, in particular the British reserves, became aware of a peculiar smell and stinging eyes. Then, as the German artillery fire stopped, masses of soldiers came stumbling down the roads from the direction of Langemarck. Few could speak, many were blue in the face and others were choking. The effects of the chlorine gas were severe. Within seconds of inhaling its vapour it destroyed the victim's respiratory organs bringing on choking attacks. Soldiers choked, their lungs burned and they slowly died as the gas cloud turned everything a sickly green. Thirty parts chlorine to a million parts air creates an irritant causing harsh coughing. On this day the Germans used one thousand parts chlorine to a million parts air and this proved lethal, caustically stripping the lining from the lungs and causing the victims to drown in their own fluids.[17] A more horrific description is provided by Lance-Sergeant Elmer Cotton:

> It produces a flooding in the lungs...a splitting headache, terrific thirst and to drink water is instant death, a knife-edge pain in the lungs and the coughing-up of a greenish froth off the stomach and lungs, ending finally in insensibility and death. The colour of the skin turns a greenish-black and yellow, the tongue protrudes and the eyes assume a glassy stare.[18]

In the front-line trenches where the gas was thicker soldiers had no time to run and not many survived. Rolling over the trenches the gas had overwhelmed them so quickly that men collapsed at once. Lying choking, gasping for air at the bottom of deep trenches where the gas had settled and clung thickest, they suffocated to death in minutes. In the reserve and rear trenches the air was heavy with fear and panic of the unknown. It was obvious something very serious had happened and at about 1900 hours the guns of the French divisional artillery in the sector ominously ceased fire. An eerie silence fell over the area. Shortage of troops had restricted the breadth of the German attack, but their hopes that surprise and the introduction of the new weapon would enhance its chances were fulfilled; a 4-mile gap had been created in the Allied line.[19] After half-an-hour, German troops, equipped with cotton wadding tied over their faces – a primitive form of protective mask – cautiously advanced into the breech created by the first discharge of chlorine gas on the battlefield. When the German troops who reached their objective saw the havoc their gas had wrought they refused to proceed any further that night.[20] Indeed, the commander of the German forces noted in his memoirs that, 'I must acknowledge that the plan of poisoning the enemy with gas just as they were rats sickened me as it would any decent soldier: it disgusted me.'[21] On 24 April, two days later, the Germans conducted a second chlorine gas attack at Ypres, this time against Canadian troops and indeed, although they discharged gas a further four times throughout May, the element of surprise had been lost. The Allied troops were now equipped with their own primitive masks and, although the defenders suffered severe losses, the Germans could gain no more than a few hundred yards beyond the forward limit of their first attack.[22]

The German High Command, surprised as its opponents at the success of the new weapon, had no reserves to now exploit its unexpected advantage and possible success. Thus, one of the war's greatest tactical surprises represented a squandered opportunity for the Germans. Had the gas attacks been performed on a larger scale and followed up, they could have decisively changed the course of the war. In practice the new chemical weapons just made the stalemate even more miserable. Ironically, the use of gas should not have been a surprise to the Allied troops, as captured German soldiers had revealed the imminent use of gas on the Western Front. On 20 March 1915 some prisoners had been captured and under interrogation had given extensive details of the plan and of the placing of cylinders in the trenches.[23] The idea was regarded as being so fantastic that a half-hearted report was eventually filtered up the chain of command but arrived long after the reporting division had been posted to a new area. It was, however, published in the *Army Bulletin* but circulated only in the Artois area, 100 miles away. A week before the attack, on 15 April, a German deserter, Private Auguste Jaeger of the 234 Reserve Infantry Regiment, revealed the exact area which was to be attacked, near Langemarck, and, as evidence, showed the respirator German infantry had already been issued with.[24] British and French intelligence officers concluded, although Jaeger's evidence was convincing, it was a little too convincing. The German had been too easily captured, perhaps he had been sent on purpose to deliberately mislead. After all, the use of poison gas was strictly proscribed in the Hague Convention, and all civilised nations, including Germany, had signed it. However, the Germans were not above employing devious tactics, particularly in the field of propaganda, and indeed on 17 April the German newspapers carried a story that the British had committed the crime of using poison gas against defenceless German troops, contravening not only the rules of war but the unwritten laws of civilisation.[25] It was, of course, a cover story designed to justify the fact that the Germans themselves were planning to use such a weapon in an attack which they hoped would be seen simply as retaliation. All these warnings might just as well have never been given for the heed that was paid to them. Consequently, the soldiers in the trenches had no warning.

Just what the effects of the chlorine gas attacks at Ypres were in terms of gas casualties is uncertain. A British author writing in 1919 stated there were at least 5000 dead, with many times that number wounded.[26] A few years later, a German writer gave figures of 15,000 casualties and 5000 dead,[27] but in 1934 he withdrew the figures saying that for propaganda reasons the Allies had quintupled their casualty figures.[28] The French, who suffered most from gas at Ypres, do not appear to have published their casualty figures: probably the necessary records could not be made. In the absence of these the most reliable indications are the statistics contained in the official British medical history of the war, based on the war diaries of the medical units in the Ypres area. About 7000 gas casualties passed through the field ambulances and casualty-clearing stations; 350 of these subsequently died.[29] It is not clear whether these statistics refer to British or French troops, or both. They do not include gas casualties taken prisoner by the Germans or those who were not admitted to medical aid posts, whether because they died or recovered before reaching them. As to prisoners of war, a German source quotes a figure of 200 Allied gas casualties admitted to German hospitals of which 12 subsequently died.[30] It is impossible to say how many failed to reach medical aid stations. A British authority estimates that the figure of 7000 should be expanded by a further 3000, mostly dead.[31] Part of the dead would have fallen on the ground soon to be occupied by the Germans, and although one German writer states that an army doctor visiting captured French trenches on 23 April could not discover a single gassed corpse, this is a rather partisan account and it seems unlikely that the French retired quite as quickly as this would imply.[32]

The German use of chlorine gas provoked immediate widespread condemnation and indignation which, though fanned by the newspapers, was deep and abiding among the soldiers. War had been envisioned as a great game, where the young men of Europe would come into their own on the battlefield and the decadence of modern society would be purged. Such notions had, of course, been blasted out of the soldiers at the front by the more than one million dead at this point in the war, but gas was seen by many as simply too much. Certainly it damaged German relations with the neutral powers, including the United States. As the *New York Tribune* reported, 'The nature of the gases carried by the German asphyxiating shells remain a mystery... That such devices might be used in war has been known for a long time, but the positive prohibitions of the Hague Conference have prevented the more civilised nations of Europe from... experiments in this line';[33] and the following day, 'The gaseous vapour which the Germans used... contrary to the rules of the Hague Convention, introduces a new element into warfare.'[34] In Britain, in the House of Lords, Lord Kitchener stated, 'The Germans... used asphyxiating and deleterious gases when their attack, according to the rules of war, might otherwise have failed... I would remind you Germany was a signatory to the following article in the Hague Convention: "The Contracting Powers agree to abstain from the use of projectiles the object of which is the diffusion of asphyxiating or deleterious gases." '[35] Lord Kitchener was indeed moved from his usual restraint to passionate anger, writing to the French that the attack was 'contrary to the rules and usages of war'. The attack was certainly contrary to usage, but the wording of the Hague Convention did not explicitly condemn it. The action then may well be held to be contrary to the spirit, but a legal document must be interpreted according to the text. The attack had one clear benefit in Germany however, for it brought to an end German hesitancy (and disagreement) over its use. Indeed, German newspapers were so enthusiastic over the effectiveness of poison gas that some even claimed that chemical weapons were more humane than bullets and shells.[36] From this point onwards the use of chemical weapons continued to escalate for the remainder of the war.

With the battlefield at Ypres now stabilised, the British and French had to decide whether or not to retaliate in kind. Faced with the German's obvious technical advantage, the Allies at first hesitated to retaliate for fear of inviting the expansion of chemical warfare, but when Sir John French reported that a lack of offensive gas capability would seriously impair the morale of his troops, the British cabinet gave its approval for the use of chemical warfare agents.[37] The French government soon followed suit for the same reason. However, although it was relatively simple to produce the poison gas, it was much harder to employ it effectively in a war situation and there was simply not enough time to train men in this novel form of warfare. As was later the case with the tank, it could be argued gas technology was misused through an inability to exploit it to its full potential. The British, however, were the first to attempt to respond by raising Special Gas Companies in the wake of the German's April attacks.[38] The entire unit initially consisted of approximately 1400 men drawn from those who had worked in the chemical industry or had been chemists in civilian life, and operating under Major Charles Foulkes (promoted to Lieutenant-Colonel on accepting the command). Foulkes was energetic and capable and quickly implemented schemes for gas defence and offence. For example, a Canadian soldier reported to Foulkes that he had observed a German soldier pulling a bag over his head during a gas attack. This comment resulted in Foulkes ordering respirator tests and the consequent British development of the Hypo-Helmet, primitive gas masks made of flannel that was chemically impregnated to neutralise chlorine with eyepieces made out of celluloid. By June 1915 two million Hypo-Helmets had been issued to Allied troops.[39] These helmets were better than nothing but could not resist an extended gas attack. But, given enough gas, any filter would eventually become saturated and ineffective.

The first draft of soldiers to the Special Gas Companies reached Foulkes at St Omer in France on 18 July 1915 and more appeared on 21 July raising the total contingent to about 2000 men. An experimental ground was set up at Helfaut and training commenced in meteorology, theory and the practicalities of gas cylinder deployment.[40] It was also stressed to the men that they were not allowed to refer to the word 'gas' in their operations, such was the stigma attached to its use. Indeed, they were warned that the use of the word 'gas' brought severe punishment; hence they referred to their gas cylinders as 'accessories'. Despite great problems in Britain in the production of chlorine and the supply of cylinders, the stockpile grew. The first British attack was now fixed for 15 September at Loos. On 4 September 1915 the Special Gas Companies that had finished training moved up to the front and began emplacing the 5500 cylinders to be used to discharge the chlorine; all wore brassards of pink, white and green, colours that were perpetuated at Porton Down Chemical Warfare Establishment until 1979 in the mess tie.[41]

At a conference on 6 September, General Haig explained to his Corps commanders the role that the gas was expected to play in the attack which had now been rescheduled for 25 September. Haig blithely spoke of the gas being carried on the wind in front of the assaulting divisions and creating a panic in the German ranks and hopefully, he stressed, that would incapacitate them for a prolonged period. However, in reality, Haig's suggestion of a panic among the enemy when confronted with gas was little more than a fantasy, for opposing the British troops at Loos were seasoned German troops equipped with respirators. Under the circumstances there was no justification for expecting the Germans to panic, and indeed, without gas masks, the Canadians had held their lines against a German gas attack at Ypres in April. With hindsight, it is easy to say that perhaps too much was being expected of this new, unreliable weapon.

Dry weather had been prevalent in the middle fortnight of September, but when the attack began wet and misty weather had set in. Yet the levels of precipitation were not as vital to the success of the attack as the direction of the wind. During the evening prior to the attack the winds had died and the following morning the British commander, General Sir Douglas Haig, made a controversial decision to proceed with the attack despite uncertainty as to whether or not the slight breeze that rose on the morning would continue to blow towards the German lines. The decision to attack seems to have simply been based on Haig's appraisal of how the wind was affecting the smoke off a cigarette. Had he watched the cigarette smoke more intently, however, he would have seen that the smoke was drifting towards the German lines but then stopping and, if anything, coming back.

At 0550 hours the guns began to fire and a mixture of smoke and chlorine gas was released intermittently over a period of 40 minutes before the infantry assault began. However, releasing gas from cylinders in this manner meant that the user had to be wary of wind conditions since, as discussed, it was desirable that a light wind exist in the direction of the enemy trenches; if the wind were to turn, however, the gas would then be swept back to the Allied trenches. In parts of the British line that morning this is precisely what transpired. The wind shifted and quantities of the smoke and gas were blown back into the British trenches engulfing the troops waiting to attack. It has been estimated that the British suffered more casualties that morning than German.[42] Nevertheless, the Germans were taken by surprise as the war diary of the German Sixth Army, the unit that bore the brunt of the attack, reveals.[43] The gas in some cases caused only momentary confusion, but in other cases entire units lost their ability to resist the British infantry follow-up attack. The German protective mask broke down as the gas lingered. The chlorine also caused rifles, machine guns and even artillery breechblocks to jam. However, the most effective result of the gas was that it rendered German officers incapable of shouting commands loud enough to be heard through their masks. IV Corps suffered the worst British casualties on the left wing of the attack. On this front the wind was blowing from the south-west which meant that the soldiers were not only hit by their own gas but by that released by units to their right. Within a few minutes 300 men were down with gas poisoning and it also became apparent that the British respirators were not working properly and many more men were gassed even though they were wearing them.[44] Ironically, while the British chemical weapon was inflicting heavy British casualties, traditional bayonet and bullet were helping the British soldiers, who had crossed no-man's-land safely, to achieve considerable success on the first day of the battle. Gas had played no part in this success other than its potential to conceal the advance of the British troops across no-man's-land. It is clear that Haig's decision to release the gas in the circumstances that prevailed on 25 September was unjustifiable and the heavy casualties of over 2500 men could have been avoided. Although the casualty numbers are arguable, there is little doubt that the British retaliatory exercise proved a failure and this was for three reasons. The first was the decision to proceed with the attack despite the unfavourable wind conditions; after all, the

Germans waited a month for a favourable wind before launching their first gas attack at Ypres. Secondly, the British artillery was hampered in providing support because it lacked sufficient shells,[45] and thirdly, there were no reserve divisions to exploit a breakthrough, as indeed had been the case in the German attack at Ypres in April. In his report Sir John French acknowledged that although the attack failed to penetrate the German lines, the 'gas attack met with a marked success, and produced a demoralising effect in some opposing units'.[46] However, on a positive note, Britain's retaliatory capability had been demonstrated within five months of the first German use of gas on the Western Front. More importantly, by this action the major belligerents had accepted and expanded the use of chemicals as weapons of war. The ensuing chemical war proved to be one of experimentation with gases and with defensive and offensive equipment. As tactical doctrine evolved to reflect technological changes, the availability of gases and the imagination of commanders became the only limits to the employment of this new weapon.

The development and use of phosgene gas soon followed the use of chlorine gas. Phosgene as a weapon was more potent than chlorine in that while the latter was potentially deadly it caused the victim to violently cough and choke; phosgene caused much less coughing with the result that more of it was inhaled. Phosgene worked by causing fluid to enter the lungs and thereby preventing oxygen from reaching the blood. Additionally, phosgene often had a delayed effect and apparently healthy soldiers were taken down with phosgene gas poisoning up to 48 hours after inhalation. Now the gases were mixed with deadly outcomes; the so-called White Star mixture of phosgene and chlorine was commonly used on the Somme (1916): the chlorine content supplying the necessary vapour with which to carry the phosgene over larger distances, thereby causing greater casualties.

Remaining constantly ahead in terms of chemical warfare development, Germany unveiled an enhanced form of gas weaponry against the Russians at Riga in September 1916: mustard gas contained in artillery shells.[47] The serious blisters it caused, both externally and internally, brought on several hours after exposure, distinguished mustard gas, an almost odourless chemical, and consequently protection against mustard gas proved more difficult than against either chlorine or phosgene gas. The Germans first used mustard gas against the British on the night of 12–13 July 1917 at Ypres and the attack caught the Allies completely by surprise. During the attack British infantry saw the gas shells explode but were unable to see, smell or taste any agent, nor did they feel any immediate effects. The soldiers concluded the Germans were trying to trick them and did not put on their masks. After several hours the soldiers began to complain of pain in their eyes, throats and lungs and later blisters appeared on the exposed skin.[48] The German use of mustard gas caused British casualties, which had been declining, to increase markedly.[49] However, the use of mustard gas had mixed benefits. While inflicting serious injury on the enemy the chemical remained potent in the soil for weeks after release making capture of infected trenches a dangerous undertaking and, additionally, mustard gas would freeze in the winter and still be toxic when it thawed in the spring. Even today French citizens are still occasionally suffering chemical burns from stumbling across ancient mustard shells ploughed up on old battle-fields. Perhaps in one of the ultimate ironies of the history of chemical warfare, the British had tested mustard gas during the summer of 1916, but the developers had been unable to convince the military of its utility. Nevertheless, as with chlorine and phosgene before it, the Allies promptly reciprocated by copying the German's use of mustard gas. By 1918 the use of poison gases had become widespread, particularly on the Western Front, and indeed, had the war continued into 1919 both sides had planned on inserting poison gases into 30–50 per cent of all manufactured shells.[50]

As the role of chemical warfare became more important, the demands for chemical agents rapidly increased. Ever-larger quantities of well-known chemicals were required and supplies were needed of some chemicals which previously had been made in only small amounts in the laboratory. The chemical industry was called upon to meet all these demands not only quantitatively but also quickly. Chlorine was undoubtedly the most important chemical connected with the manufacture of chemical warfare agents. Not only was chlorine itself used as a chemical warfare agent, but it was also needed for the manufacture of most of the other agents used during the First World War. Another chemical needed in quantity was phosgene. Prior to 1915, no bulk manufacture of this compound had been undertaken and yet by 1918 almost 700 tons of phosgene a year were manufactured in Britain alone.[51] However, probably the most difficult problem that faced the chemical industry during these years was to satisfy the demand for mustard gas. While in Britain the inorganic side of the industry was highly developed, the manufacture of most organic chemicals was in a backward condition due to the monopoly Germany had been able to secure through its dyestuffs industry. Consequently, Germany was well placed to manufacture mustard gas on a very large scale. In Germany substantial quantities of thiodiglycol were being produced for making dyestuffs, and good quality mustard gas could be made from this by reacting it with hydrogen chloride. On the other hand, there was no commercial source of thiodiglycol in the United Kingdom and consequently only very small amounts of mustard gas were produced in 1917 using this process. British scientists sought another method and re-examined the process by which mustard gas was first obtained by Depretz in 1822 and Guthrie in 1859, by the interaction of sulphur monochloride and ethylene. Working fast, against time, they found the optimum conditions for the reaction so that a good yield of the product could be obtained and rapidly put the method into production. The British used mustard gas for the first time in 1918, and by the Armistice 500 tons of the agent had been made by this process.[52]

In addition to production problems, the chemical industry was required to study the manufacture of a variety of other chemical warfare agents during the First World War. Among these were ethyl iodoacetate, hydrogen cyanide and toxic arsenic compounds. Large quantities of these toxic chemicals were wanted quickly and there was not sufficient time to make a careful examination of the methods of preparation. It was impossible to develop in an orderly fashion through pilot plant and semi-technical stages the process that would have been best suited to bulk production. In many cases it was necessary to translate laboratory practice direct to full-scale plant with the result that innumerable difficulties were encountered as regards both the process and the nature of the final product. Thus, makeshift methods of manufacture, often highly dangerous to the workers, came into operation.

During the First World War, chemists on both sides investigated over 3000 chemical substances for potential use as weapons. Of these only 30 were used in combat and only 12 achieved the desired military results.[53] Other types of gases produced by the belligerents included bromine and chloropicrin, and the French Army occasionally made use of a primitive nerve gas obtained from prussic acid. However, the three forms of gas already discussed – chlorine, phosgene and mustard – remained the most widely used. Table 2.1 illustrates that the German Army ended the war as the heaviest user of gas: in fact, it is suggested German use reached 68,000 tons; the French utilised 36,000 tons and the British 25,000 tons.[54]

At the same time as they experimented with more lethal chemical agents, both sides worked to develop more effective methods of agent delivery. Gas cloud attacks relied on the wind; in the absence of wind or if the wind blew from the wrong direction, gas cylinders were useless.

Table 2.1 Production of chemical warfare agents during
the First World War (in tons)

	Chlorine	Phosgene	Mustard
Germany	58,100	18,100	7,600
France	12,500	5,700	2,000
Britain	20,800	1,400	500
United States	2,400	1,400	900

Note: L.F. Haber, *The Poisonous Cloud: Chemical Warfare in the First
World War*, Oxford: Clarendon Press (1986), p. 170.

Despite these problems the British relied on cylinders as a delivery
method until the end of the war but several factors influenced the British
decision to continue using them. First, the prevailing winds on the
Western Front favoured Allied gas clouds; secondly, the British suffered
from a chronic shortage of shells and were reluctant to convert the
production of high-explosive shells to the production of gas shells; and
thirdly, British intelligence reports indicated a dense cloud attack was
effective in producing mass casualties.[55]

Because the prevailing winds in western Europe blew from west to
east, the German Army began to place increasing reliance on gas-filled
shells that detonated beyond Allied lines and whose contents could
then drift back over enemy trenches. The Germans were further
encouraged to use gas shells by the results of an attack on the night of
22–23 June 1916. Ten thousand shells containing the lung irritant
phosgene fell on the French forces near the fortress of Verdun, and
German batteries adjacent to this sector added thousands of rounds of
a lachrymatory gas.[56] The attack caused over 1600 casualties.[57] From
this point onwards the German High Command directed that all artillery
units should fire gas shells and, indeed, by the end of the war, gas shells
comprised 50 per cent of German artillery stocks.[58]

The British faced a constant artillery shell production shortage and
supplemented their use of gas cylinders with the 4-in. Stokes Mortar,
first fielded in 1915 at Loos. The weapon, designed specifically to fire
gas and thermite shells, had a payload three times as large as could be
fired from the standard 3-in. mortar. This device represented the first
use of projectiles filled with lethal chemicals in the First World War and
indeed, arguably, the first direct contravention of the Hague Declaration
(1899). Accordingly, the Germans produced chemical agent-filled projectiles for 77-mm, 105-mm and 150-mm artillery pieces, and the French

produced chemical agent projectiles for their 75-mm rapid firing guns. Throughout the First World War there were many advances in the offensive use of chemical weapons and perhaps one of the greatest innovations was the Livens projector, a large-scale mortar developed by Lieutenant W.H. Livens, which was capable of delivering large amounts of chemical warfare agents. The original projectors were designed to throw incendiaries (not high explosives) and were constructed from 12-gallon oil drums (12 in. across × 20 in. high). After experimenting with various discharge methods, for example fuel and fuel ignition, Livens was ready with his new projectors which took him only one week from invention to destructive weapon. The new weapon was used for the first time, experimentally, in the Battle of the Somme in October 1916 and for the first time on a large scale supporting the Canadian attack on Vimy Ridge near Arras in April 1917. Interestingly, in another tactical change, prior to the Battle of Vimy Ridge in April 1917 the Canadians had launched at least 55 trench raids on the opposing German forces.[59] The aim of these raids was fourfold: to hit back at the enemy, exert some control over no-man's-land, gather information on what units were opposite, and to look for new fortifications and gas canister emplacements. The Germans reported that the density of gas delivered by the Livens Projector equalled that of a gas cloud. Captured German documents claimed that the Livens Projector was a deadly weapon because it not only developed a dense concentration of gas similar to the ones created by the cylinders, but like artillery, its impact came as a surprise.[60]

Increased casualties resulting from the British Livens Projector attacks prompted the Germans to develop a similar weapon. However, time restraints and a lack of industrial capacity for increased steel production forced them to retool their obsolete 18-cm heavy mortars; ultimately these tubes were able to fire a projectile containing four gallons of a chemical agent. In August 1918 they introduced a rifled projector, 16 cm in diameter, that increased the range of the device to 3500 m. The shells contained 13 lb of chemical agent and 5 lb of pumice. The pumice kept the chemical agent from being flung into the air upon explosion and it also made the agent, usually phosgene, more persistent. Indeed, in one instance, the gas reportedly lingered for one and half-hours.[61] Yet, impressive as these results were, the Germans despite their efforts continued to lag behind the British in the tactical use of chemical warfare delivery systems.

Nevertheless, from 1915 to 1918 the Germans still held the initiative in most areas of chemical warfare. They did this through the introduction of new agents that allowed them to direct more systematic thought to the question of how the employment of gas might alter a tactical position.

They were, for example, the first to use gas as an addition to manoeuvre in support of an infantry attack. The Allies struggled to keep up with such offensive doctrine and they had to contend first with the development of effective defensive measures to counter such German initiatives. Only after developing counter-measures could the Allies then plan their use of a new chemical agent or a new delivery system. This lag was evident in the case of the two most effective chemical agents used in the First World War, phosgene and mustard gas. The Germans introduced phosgene six months before the Allies were able to employ it, and mustard a year ahead of their foe.[62] The Allies had to adopt immediate defensive measures, such as effective mask filters and protective suits, before they could turn to the development of tactical doctrine.

British gas doctrine, when the circumstances did permit its development, was driven in part by a shortage of artillery shells that prohibited the British Army from mounting an artillery gas attack until the summer of 1916. In the meantime, the British convinced themselves that chemicals released from cylinders or projectors could most effectively be used to obtain the highest possible concentration of a chemical agent in a specific area.[63] The consequences of this doctrine were twofold: First, it prevented the British from employing gas to support mobile or open warfare, and secondly, it limited the use of chemical agents to the more restricted roles of attrition and harassment. In the case of harassment, the British High Command, relying on its intelligence reports, would indicate what German units it wished to weaken or demoralise. German divisions recently transferred from the Eastern Front were prime targets because of their ignorance of defensive measures against gas warfare. The British sought out units they expected to be transferred to the main battle-fronts, in other words, Somme and Ypres, and tried to weaken them physically and psychologically before they were deployed. On at least one occasion a gas operation was postponed to await the arrival of a particular division. The 1st Bavarian Regiment, for example, was gassed fifteen times; the 1st Guards Regiment, twelve times in six months; the 10th Bavarian Regiment, ten times in five months and the 9th Bavarian Regiment fourteen times from 28 June 1916 to 1 August 1917.[64] The effects could be devastating to the morale of the gassed units and those units around them. A German diary recorded, 'We have again had many casualties through gas poisoning. I can't think of anything worse; wherever one goes, one must take one's gas mask with one, and it will soon be more necessary than a rifle. Things are dreadful here.'[65]

The British ultimately developed a tactical doctrine for the use of gas shells.[66] This doctrine set three methods for inflicting enemy gas casualties. The first method was by surprise attack, the second was to use gas shells to try to exhaust the enemy by random fire over a period of many hours, but in most instances the British believed this attrition method not worth the effort because very few casualties were produced. The third method was an attempt to penetrate the enemy's gas masks with new agents such as chloropicrin, which when fired in high concentration in a specific area seeped into the masks and created intolerable eye irritation, coughing, vomiting and inflammation of the respiratory tract. Enemy soldiers, forced to remove their masks, were then subjected to a shelling with lethal phosgene.[67]

However, the Germans still had the technological advantage and that gave them the ability to introduce new gases before the Allies and indeed they gave much thought to the tactical deployment of chemical weapons, and in this respect they reached a high degree of sophistication. After abandoning gas cloud attacks, the Germans increased their use of gas shells. On the Western Front in 1916 they fired some 2000 tear-gas shells at an extensive French trench system near Verdun. This massive surprise bombardment resulted in the capture of 2400 Frenchmen who, after being temporarily blinded by the tear gas were surrounded by German troops wearing goggles, but no masks.[68]

The Germans introduced other agents to the battlefield for specific tactical purposes. For example, in May 1916 they fired a shell filled with diphosgene, a lung irritant, ensuring panic and temporary inactivity by the enemy. Later, as an indication of the sophistication of gas shells, they subdivided the shell mix, first by a mix of 75 per cent phosgene to 25 per cent diphosgene. Then, in July 1917, three different percentages of phosgene, diphosgene and diphenylchlorosine were introduced.[69] The introduction of mustard gas, however, gave the Germans the initiative in chemical warfare, which they held until the end of the war. The Germans found that gas persisted even longer when an agent and a small amount of high explosive were placed in one shell. The effect of the high explosive, when used in proper amount, was to spread the agent over a wider area and keep it airborne longer.

With this knowledge the Germans changed their gas doctrine from attacking a particular target to gassing large areas for extended periods of time. The key figure in the expansion of German gas shell doctrine was Lieutenant-Colonel Georg Bruchmüller, known as *Durchbruk* (breakthrough) and considered an artillery genius because of his success on the battlefield. Bruchmüller was a great believer in the efficiency of gas shells and his tactical ideas were incorporated in the December 1917 edition of the *German Manual for the Employment of Gas Shells*.[70]

In one of the final Allied chemical assaults of the First World War the British fired mustard gas into German positions at Wervick in Belgium. One of the injured was a corporal by the name of Adolf Hitler who was evacuated back to Germany burned and temporarily blinded. As a result of that experience, Hitler developed a distaste for the use of poison gas on the battlefield. The experience of this German corporal would, in turn, shape the events of another world war 20 years later.

Assessment of the impact of the use of gas in the First World War on the Western, Eastern and Italian Fronts is difficult. Analysis of casualty figures is doomed to failure because of a contemporary lack of definition and classification. Gas casualty estimates by several national sources exceed a million but elements of uncertainty exist on the precise cause of death or major source of injury in those who were both gassed and wounded. Also comparison of gas and other battlefield injuries shows vast swings in the proportions on different fronts in different years (Table 2.2).

Table 2.2 includes both fatal and non-fatal casualties and these figures are based on admissions to medical units in France. They do not, therefore, include chemical warfare casualties captured by the Germans, minor chemical warfare casualties who were returned to their units from field ambulances without treatment, chemical warfare casualties

Table 2.2 Chemical warfare casualties of the BEF in France during the First World War

	Total battle casualties	Battle dead as % of all battle casualties	Total chemical warfare casualties	Chemical warfare casualties as % of all battle casualties	Total chemical warfare dead	Chemical warfare dead as % of all chemical warfare casualties
1915	304,406	26	12,792	4.2	307	2.4
1916	636,146	27	6,698	1.1	1,123	17.0
1917	727,022	29	52,452	7.2	1,796	3.4
1918	768,603	25	113,764	15.0	2,673	2.4
1915–1918	2,436,177	27	185,706	7.6	5,899	3.2

Source: T.J. Mitchell and G.M. Smith, *Official History of the Great War: Medical Services; Casualties and Medical Statistics of the Great War*, London (1931). (Figures for 1915 refer to British casualties only, while those for later years include British Dominion casualties as well. The 1915 figures therefore do not include the heavy Canadian chemical warfare casualties during the Second Battle of Ypres.)

who died on the battlefield and non-fatal chemical warfare casualties killed by other weapons. Indeed, such figures confirm the view of L.F. Haber who suggests that data on casualties, the cost of chemical munitions and their uses and attempts at cost-effective analysis lead to a dead end.[71] Similar casualty lists are available for the German forces, but only those compiled from casualties brought into Allied medical centres and therefore are not a true reflection of German chemical warfare casualties.[72]

The story of gas in the First World War, on all sides, is one of experiment and imitation conducted on a background of uncertainty and hurriedly assembled arrangements for development, production and use. Unfamiliarity and a lack of confidence, both to exploit gas warfare to the fullest and to seize the initiatives revealed after successful attacks, complicated this. The real utility of gas in the First World War cannot be determined. It brought no great victories, yet it had an obvious military impact. Those who remained unprotected were vulnerable to the extent that all armies perceived the need to have high levels of gas protection and, where possible, to develop and maintain the ability to retaliate in kind. The advent of mustard gas emphasised such needs. The greater impact was perhaps to evoke a level of public horror subsequently reflected in political concern of an enormity sufficient to press for the very legality of chemical warfare to be considered and for arms control measures to be applied. A major factor influencing both public and official minds was the apparent future vulnerability of the civil population to the use of aerial chemical bombs in war.

In Britain, in the military there were conflicting views on the matter of chemical weapons. On the one hand, some senior officers were all for the urgent development of chemical weapons as an essential in future wars, on the other hand, others were less convinced although conscious of the possible forfeits of non-possession. Some even thought that problems of defence were so great that no consideration should be given to any use of chemical warfare in future wars.[73] This view probably arose because of fears that military scientists might de-stabilise conventional military doctrine to the extent that the conventional means of war familiar to the professional soldier would be subsumed in the still unconventional chemical warfare. Equally, those with personal experience of gas in the trenches had other and humanitarian reasons for obstructing further developments in these means of war: emotive assessments were undoubtedly made. The supporters for the further development of chemical weapons pressed the alternative humanitarian view that short-term incapacitation from chemicals was the rule, rather than death, and that, apart from the deaths associated with the early cloud attacks against unprotected or poorly protected troops, chemical warfare had not resulted in a large proportion of deaths.[74] Certainly this view was strengthened by J.C.S. Haldane in 1925 who stated: 'Of the 150,000 British mustard gas casualties less than 4,000 [1 in 200] became permanently unfit.'[75]

It has been estimated that among British forces the number of gas casualties from May 1915 to July 1917 amounted to 9 per cent of the total, but of this total only around 3 per cent were fatal (Table 2.3). However, from July 1917 to the end of the war the use of mustard gas produced higher casualties with greater fatalities.[76] Mustard gas was named the 'King of Gases' for this destructive potential. Mustard gas victims often led highly debilitating lives thereafter with many unable to seek employment once they were discharged from the army. Additionally, because of their weakened state many succumbed to the Spanish Flu epidemic that swept across Europe from late 1918, perhaps inflating casualty figures for this pandemic.

In large part the lower casualty figures from May 1915 to July 1917 can be attributed to the increasing effectiveness of methods developed to protect against poison gas but, in reality, it could be argued that gas never quite became the weapon that turned the tide of war as it was predicted to be. It is difficult to find a definitive figure for the numbers of men injured and killed by chemical warfare agents during the First World War (Table 2.4). British casualties alone can be estimated at 85,000 injured and 8700 dead.[77] Another historian gives a total figure of 1,296,853 casualties produced by 125,000 tons of chemical warfare agents used by all combatants,[78] but it is known that in many cases the official figures underestimate the number of casualties.

Furthermore, it is unclear to what degree the official figures include individuals who were injured in gas attacks but who developed serious symptoms only after the war. Given the general estimate of 10

Table 2.3   British gas casualties (Western Front)

Date	Gas	Fatal	Non-Fatal
April to May 1915	Chlorine	350	7,000
May 1915 to June 1916	Chlorine	0	0
December 1916 to July 1917	Phosgene	532	8,806
July 1917 to November 1918	Mustard	4,086	160,526

Note: Compiled from PRO, WO 32/5951, casualties caused by poison gas in British forces 1915–1918 (1918–1919).

Table 2.4   Gas casualty figures for each belligerent during the First World War

	Total casualties from chemical warfare agents	Fatal casualties from chemical warfare agents
Germany	200,000	9,000
France	190,000	8,000
British Empire	189,000	8,100
Austro-Hungary	100,000	3,000
Italy	60,000	4,600
Russia	475,000	56,000
USA	73,000	1,500
Belgium	73,000	1,500
Romania/ Bulgaria	10,000	1,000
Total	1,297,000	91,000

Source: A.M. Prentiss, *Chemicals in War*, New York (1937).
Note: The figures given in this table are very rough approximations. Only in the cases of the UK and the USA are reasonably adequate casualty statistics available. For the other belligerents, the figures given are those estimated by Colonel Prentiss of the US Army Chemical Warfare Service after his careful study of all available material. His treatise, op. cit. *Chemicals in War*, should be consulted for further information about his estimates.

million battle deaths from the First World War, it is arguable whether chemical warfare was more or less horrific than the other methods of conducting war.

The historical developments of crucial equipment of modern warfare often have their roots in peacetime scientific applications, as indeed did the development of poison gas, being an offshoot of the dyestuffs industry. New scientific enquiry and the extension and refinement of existing technology have been the foundations of military development since the Stone Age. 'You can't say that civilisation don't advance; for in every war they kill you in a new way.'[79] This sentiment, although obviously intended as a humorous statement, does highlight the concept of military application and utilisation of existing and developing technologies. As history has shown on many occasions the defensive implications or military necessity for a given technology often serves as the engine that drives its development.

When examining the technical and scientific evolution of the respirator, the question of original invention inevitably arises. As Galarraga points out, while it is very likely that human beings used makeshift masks for thousands of years to protect their eyes, mouths and respiratory systems from smoke and dust, the first detailed description of a protective mask is usually credited to Leonardo da Vinci in the early 1500s.[80] Ironically, da Vinci's mask was proposed as a counter for offensive technology that he had been stimulated to develop, the precursor to the modern chemical shell. Like most of da Vinci's work, however, his concepts of chemical warfare and protection were far ahead of their time and the first real evidence we have of a protective mask being employed was in 1665 during the Great Plague.[81] This mask covered the eyes, nose and mouth of the wearer and, although primitive by today's standards, this invention clearly demonstrates a historical attempt to design a protective device specifically for shielding the user's respiratory system. This mask, not based on the soundest of designs, did not lend itself to military application but mainly because there was no call for respirators in war fighting at this time. The first serious attempts at developing respirators occurred between 1849 and 1910 but these designs were targeted towards fire fighters and chemical industry workers. However, it was during this time that the first gas mask, to use wood charcoal rather than dampened cloth as a filter, was developed. It is clear then that defensive military technologies follow the development of offensive war-fighting technologies, and although the precursors of chemical warfare were established about 2000 BC their effective use on the modern industrialised battlefield had not yet been witnessed.

So, while much energy was devoted to offensive aspects of chemical warfare the protection of the soldier was considered a no less important matter. In Britain the first Anti-Gas Departments were in London at the Royal Army Medical College and later at the University of London (UCL). Following the first gas attacks, the immediate need was for the provision of gas masks (or 'respirators' as they became known) and indeed within 36 hours of the first gas attack against the French forces, 100,000 wads of cotton pads were quickly manufactured and made available. These were dipped in a solution of bicarbonate of soda and held over the face but since they were available in only limited numbers, soldiers were also advised that holding a urine-drenched cloth over their mouth would serve in an emergency to protect the respiratory system against the effects of chlorine. Clearly the wads had limited utility and the idea of an impregnated flannel helmet was conceived and goggles were also produced to compliment some devices. In the autumn of 1915 British intelligence learned of the German intention to use a new gas, phosgene. The Russians advised the British that a solution of phenate-hexamine was effective in blocking the chemical agent and as a result the British soaked their 'Hypo-Helmets' in the solution, added a device to reduce the carbon-dioxide build-up inside the mask and renamed it as 'PH Helmet'.[82] The troops called it a 'goggle-eyed bugger with a tit'.[83] Although the PH Helmet successfully blocked phosgene, it had serious drawbacks: it was hot, stuffy and emitted an unpleasant odour; it also offered little protection against dense concentrations of lachrymatory agents. A more efficient and comfortable concept of an impervious facepiece with eyepieces and the essential gas absorbents and filters incorporated in an attached container soon arose. Such a mask, it was argued, would have the merit that any necessary specific absorbents for new gases could be added as an additional layer in the container, a method not feasible with a fabric helmet. The first 'Large Box' respirator was issued in August 1916 and had a container holding soda lime-permanganate granules, a facepiece of proofed fabric, mouthpiece, nose-clip and separate goggles. The facepiece was connected to the 'box' by a rubber tube. As with most respirators, numerous continuous improvements and modifications occurred, resulting in the eventual emergence of the 'Small Box' respirator in the later months of 1916. This respirator continued in use until the end of the war.

However, while German troops were the first to initiate chemical gas attacks, the German chemical corps still relied on relatively unsophisticated protective measures. Indeed, it was not until the Allies were on the brink of chemical retaliation that German troops received effective gas masks of their own as standard issue equipment. The most common in early usage was the *Gummimaske*, which was made of fabric with a thin layer of rubber connecting the filter directly to the facepiece. However, the mask was considerably lighter than those of the Allies. This was later supplemented by the issue of Dräger's one-hour *Heeres Sauerstaff Schutzgerät* (HSS-Gerät) or 'Army oxygen equipment' in 1916. More than 100,000 HSS-Gerät sets were eventually issued to German troops during the war,[84] but the respirators proved insufficient for prolonged use in gas attacks and were thereafter used mainly for general engineering and rescue work, and in the underground tunnels that both the British and German soldiers were digging under each others front line in order to set explosive mines.[85] Nevertheless, there is evidence that the HSS-Gerät remained in service after the introduction of the more complex *Heeresatmar*, and indeed was seemingly still in use at the end of the Second World War, despite no longer being manufactured.[86]

Little other individual protective equipment emerged during the First World War beyond impregnated leather gloves and linseed oil-impregnated suits for occasional use by troops in areas where mustard gas had been employed. These items were not in general use however, and the war ended before the particular problems associated with the protection of the skin against mustard gas had been studied. The only other notable protective equipment to emerge was a cover for messenger pigeon baskets and some desultory studies on the use of fans to disperse gas from the trenches, studies that were soon abandoned.[87]

Gas in the First World War did not have to cause a large number of casualties to be an effective weapon. Chemical warfare placed an additional strain on every aspect of combat and there is no doubt that 'The appearance of gas on the battlefield changed the whole character of warfare.'[88] In the First World War gas was everywhere, in clothing, food and water; it corroded human skin, internal organs and even steel weapons. The smell of gas hung in the air, and the chemical environment became a reality in everyday life. Not only did men have to train constantly, but also an entire logistical network had to be established for offensive and defensive gas equipment. Despite the pervasive impact of chemical agents on the battlefield, commanders still had difficulty adjusting their thinking and planning in such a way as to make effective use of these new weapons – weapons totally different from anything they had ever been trained to use. Certainly, the experience of the Allied armies during the First World War suggests several shortcomings in the military's preparation for, and later employment of, chemical warfare.

Proper defensive equipment is a minimal requirement for the successful engagement of forces in chemical warfare. The indispensable item for the First World War soldier was his protective mask but, besides filtration of all harmful agents, the mask had to fulfil a number of other requirements to be efficient. It had to be comfortable and allow freedom of movement, full vision, easy breathing, communication and durability. The failure of all belligerents to develop a mask that could meet these requirements ultimately limited the combat effectiveness of the soldiers.

Once the First World War started and the Germans had used chlorine for the first time, the Allies should have immediately geared up for the production of war gases. Production was belatedly undertaken. The unfortunate shortage of shells restricted the Allies' ability to retaliate in kind against the Germans and this, in turn, had a demoralising effect on troops whose own positions had been liberally drenched with gas from German shells. Additionally, the Allies never found the key to effective education and training for the offensive and defensive aspects of chemical warfare. Unfortunately, Allied training in chemical warfare never reached the sophistication attained by the Germans, training that was undeniably needed to achieve the desired results. Equipment shortages and the lack of trained instructors hampered the Allies' preparation to engage in chemical warfare and they suffered needless casualties as a consequence.

Because the Allies failed to develop an effective gas warfare doctrine, the average British, French and, later, American officer never really understood the potential value of chemicals. Nor could he put aside his preconceived, perhaps erroneous, notion that chemicals were unusually inhumane weapons whose development should not be pursued. For the Allies the real inhumanity of chemical warfare in the First World War lay in the blindness of their leaders who, having ignored the real and present threat posed by gas, deployed soldiers to fight virtually unprepared in a chemical environment. Ignorance and short-sightedness exacted a high price at the front – a price that the Allies with their intellectual and technological resources should not have had to pay.

It is difficult to assess the importance of chemical warfare techniques during the First World War. Gas was one new weapon among several and, like the tank, submarine and the combat aircraft, it was employed on an increasingly large scale as the war progressed. While it was not a battle-winning weapon, there were a number of engagements on the European fronts where the outcome would have been different had gas not been used. However, it would seem doubtful if any conclusion could be reached that chemical warfare determined the overall outcome of the First World War. These considerations are now largely academic though. The facts of the matter were that some people felt gas to be an important weapon and their promotion of its use, in terms of the initiation of large development, procurement and deployment programmes, was so pervasive that by the end of the First World War gas had become a standard weapon, if not a universally popular one. Few people doubted that it would be used again in some future war, and, because its technical and military possibilities had clearly not been exhausted, it became a weapon to be taken seriously.

3
The Inter-War Years, 1919–1939

It was natural that the war should be followed by a wave of anti-war feeling. The war had done what the writing of the economists had failed to do: it had demonstrated that modern warfare brought loss on a colossal scale to the victors as well as the vanquished. The establishment of the League of Nations, and its early activities, showed a general determination to find an alternative to war for the settlement of international disputes. Nevertheless, the calls for worldwide disarmament continued and eventually legislation was passed in an effort to limit chemical weapons.

Somewhat surprisingly, following the end of the First World War, the Allied governments almost immediately seemed to forget what they had learned during the war about being prepared for future chemical warfare. The first major concerns for the chemical warfare detachments of the Allied forces then were to ensure they survived demobilisation. In both Britain and the United States cases were presented for the need for a permanent chemical warfare research establishment. In 1920 A.A. Fries proclaimed:

> Had there been a chemical warfare establishment in 1915 when the first gas attacks were made we would have been fully prepared with gases and protective masks and the army would have been trained in their use. This would have saved thousands of gas cases. The war might have easily have been shortened by six months or a year, and untold misery ... might have been saved.[1]

Fries then went on to stress that both offensive and defensive research must be conducted. Somewhat prophetically, he forecast: 'In the future gases will be found that will penetrate the best existing masks',[2] and additionally he disagreed, along with Lefebure, with the premise that treaties could prevent chemical warfare, stressing that the production of gases could be carried out anywhere, and no one would be any the wiser.[3]

Certainly, after the First World War the German chemist Fritz Haber continued his work on poison gases under the cover of 'pest control', as gas weapons had been forbidden to the Germans by Article 171 of the Treaty of Versailles in 1919.[4] During this time Haber developed an insecticide that could be used to fumigate buildings in the form of a crystalline material that released hydrogen cyanide fumes; it could also be deadly to humans in enclosed spaces. Known as 'Zyklon B' it was to take on a new significance 20 years later.

In 1919, Haber was awarded the Nobel Prize for chemistry for his work on nitrogen fixation. In his acceptance speech he did not, however, avoid the subject of chemical warfare saying 'In no future war will the military be able to ignore poison gas. It is a higher form of killing.'[5] They were hardly ignoring it; indeed, four classes of agents had been developed during the war and were now being refined in the post-war period. In the 'Asphyxiant Class', which had comprised chlorine and phosgene in the First World War, Diphosgene was developed which was similar to phosgene in composition and action but easier to handle. Also, Chloropicrin – known as 'vomiting gas' by the British, 'Aquinite' by the French and 'Klop' by the Germans – was being refined in the post-war period to be used in combination with other gases. The value of chloropicrin, despite the fact it was less effective than phosgene, was that it could penetrate gas mask filters more easily, and therefore could arguably be ultimately more effective. In the 'Blister Agent' class, mustard gas was the obvious example from the First World War but in the post-war period the original sulphur mustard was replaced by nitrogen mustard; nitrogen mustard was easier to manufacture and more persistent than sulphur mustard. It was in this class that the Americans made a significant contribution to chemical weapons in the form of a blistering agent named 'Lewisite', developed in 1918 by W. Lee Lewis of the Catholic University in Washington DC. Lewisite was similar to mustard gas in its ability to cause damage to a victim's entire body, but it was much faster acting. Lewisite was an oily liquid that ranged from clear to dark colour depending on the impurities present, but was lethal either way: pure product (clear) had little smell, but impure product (dark) smelled something like Geraniums. Lewisite was an arsenic-based (or arsenical) compound that caused a burning sensation within 15 seconds. Following its development the Americans built a huge production facility at Edgewood Arsenal, Maryland, to manufacture Lewisite in quantity. Indeed, Edgewood became the United States' chemical warfare centre for training, stockpiling, and research and development during the inter-war period. However, it was too late in the war to get it into service and, indeed, the Americans gave up its production soon after the end of the conflict. However, they had let the genie out of the bottle and, later, other nations would find Lewisite very interesting.

A family of other broad-effect irritants were also developed in the post-war period, known as 'nettle gases' as they made the victim feel as if he had been dragged through stinging nettles. The best-known of the nettle gases was 'phosgene oxide', but the name is somewhat misleading as it had no strong chemical relationship to phosgene and, of course, a much different action. Additionally, a range of 'non-lethal' or, perhaps more correctly, 'less-lethal' gases were also developed. Such substances are now known as riot control agents and they comprise in essence tear-gas agents not dissimilar to those employed in the First World War. Following the patterns set in other chemical classes, after the war new tear gases were developed including 'ortho-chlorobenzylidene malononitrile', mercifully better known as CS gas after its inventors Corson and Stoughton. Today not only is CS gas the basis of the popular self-defence spray 'Mace', but also CS remains in use by the military and police as a riot control agent.[6]

Delivery systems were also improved. As early as 1920, experiments with the barrel of the Stokes Mortar enlarged the bore to 4.2 in. in diameter, which increased the range of the mortar from 1100 to 2400 yards. By 1928 this new improved mortar became the standardised weapon for the delivery of toxic chemical agents, as well as smoke and high explosives.[7]

So, although calls were made to prohibit the use of chemical warfare at the end of the First World War, it is clear that the major countries continued to support investigations connected with this form of warfare. At this time many of the problems raised during the war needed to be studied in an organised scientific manner, since most of the previous effort had, by necessity, been of an empirical nature.

One of the most important tasks was to improve respiratory protection, particularly against toxic smokes, which had been a feature of chemical warfare in the latter years of the war. Little was known of the properties and behaviour of particular clouds and of methods which might be used to remove them from the atmosphere. It was recognised that many years of research was required in order to provide adequate understanding and knowledge of the particulates, which could then sanction the design of effective filters for inclusion in the new respirators. Towards the end of the First World War, charcoal was normally used as a filter in place of the chemical granules which had been popular in the beginning. The potentialities of charcoal were considerable, but it was still necessary to study its chemical and physical properties to ascertain methods for its production in a form suitable for use in respirator containers. Certainly, it can be argued that one of the most important achievements during this period between the wars was the considerable improvement in respirators and in the techniques for their production. Certainly in the United Kingdom, in the years immediately following the First World War, there was an extensive study of various types of activated charcoal with a view to selecting material suitable for use in respirators. The first supplies of respirator charcoal were made from carbonised coconut shells, but it was feared that in an emergency the demand for charcoal could not be met from this imported raw material, and so the use of alternatives, such as wood, peat, coal and coke, was examined. Eventually a process was developed using coke and this was used from 1923 for making very large quantities of high-grade charcoal for respirator containers.

The chemists of all nations faced further problems. First, it was necessary to study the properties of the more effective chemical agents that had been used during the war years so that more economical and safer methods of making them could be evolved. Secondly, it was necessary to synthesise and test new compounds that might provide better and novel chemical warfare agents. Quite apart from any intention to use such chemical weapons offensively, the information was necessary to evaluate the threat from chemical warfare and to develop adequate methods for defence. Perhaps one of the most important problems at this time was to provide an adequate defence against mustard gas. The introduction of mustard gas in 1917 had created many new problems. One of these was due to the fact that liquid mustard gas and its vapour attacked the body and produced casualties by the absorption of the agent through the skin. Therefore, in addition to protection for the respiratory tract it was clear that protection for the whole body was needed. Additionally, mustard gas, unlike the majority of chemical agents used in the First World War, was chemically stable and consequently could cause casualties long after deployment, constituting a hazard by giving off vapour and by contact as a liquid – even as a diluted liquid. Therefore, research focused on the detection of mustard gas on the terrain and equipment, and on neutralisation both on the body and on the ground. For neutralising mustard gas on the skin an ointment was developed. Originally this was a mixture of bleaching powder and Vaseline in a varnished tin. Several million of these tins were made in 1938 and

1939, but it was then discovered that in tropical climates the ointment was unstable and, indeed, became an irritant when applied to the skin. Consequently, for use in the tropics another cream was developed in which the active ingredient was chloramine T, a normal product of the chemical industry, this time contained in lead tubes. Another approach to the problem of protecting the skin was to develop permeable clothing chemically treated to destroy mustard gas. Substantial stocks of impregnated clothing were manufactured up to 1938 and held for use in an emergency. For early identification of mustard gas a chemical detector that would change colour on contact was developed. An oil-soluble dyestuff was incorporated into a cadmium lithopone paint that was used on vehicles and other equipment. The paint changed colour from green to red in the presence of any liquid vesicant. Subsequently, booklets of detector paper were also produced and widely issued.

There is evidence that chemical warfare continued in the years after the First World War, even if on a fairly small and quiet scale, despite all the proclamations made in the war's immediate aftermath calling for world disarmament. Writing in 1919, Victor Lefebure called not for treaties to limit and proscribe the use of chemical weapons but, in the mood of the time, for total disarmament of all nations.[8] Lefebure believed that a treaty to prohibit chemical warfare was inadequate, 'Chemical peace, guaranteed by a mere signature, is no peace at all',[9] and did not believe the world had '... learned its lesson in that sense'.[10] Lefebure then examined the origins of chemical warfare, accusing the Germans, and IG Farben in particular, and then prophesised that chemical warfare would continue until '... very definite steps are taken to suppress it'.[11] Finally, Lefebure concluded: 'Let us take a balanced view of the facts, realise the unique significance of chemical warfare ... for war and disarmament, and act accordingly.'[12]

The First World War saw the break-up of empires including the Ottoman Empire whose capital Constantinople became Istanbul. Out of the remnants of that empire a series of artificial states were created which arbitrarily divided up and threw together various peoples. Under League of Nations mandates, France acquired Syria and Lebanon and Britain procured Palestine, Jordan and Iraq.

The birth of Iraq was presided over by Winston Churchill who, at the time, was the British Secretary of State for the Colonies. Churchill had earlier promised Arabian ruler Sharif Hussein that he could install his son, Feisal, as ruler of Syria, but when the French seized Damascus, as a consolation, Churchill gave Feisal the lands formerly known as Mesopotania. Repeatedly from 1919 the population of this area, now

Iraq, rose up against the Hashemite ruler and his British patrons. In June 1920 a full-scale rebellion broke out and British garrisons were taken by surprise as the revolt spread throughout the lower Euphrates valley. In August the insurgents declared a provisional government. And yet, by February 1921 the revolt had been crushed with between 8000 and 9000 rebels killed. How was this accomplished so quickly? The answer is that it was achieved mainly through the use of air power, by the RAF bombing the rebels using incendiary weapons and poison gas.

In 1919, before the outbreak of the rebellion, the RAF asked Churchill for permission to use chemical weapons against 'recalcitrant Arabs as an experiment'.[13] Churchill, then Secretary for State for War, in turn asked experts if it would be possible to use '...some kind of asphyxiating bombs calculated to cause disablement of some kind but not death... for use in preliminary operations against turbulent tribes'. Churchill then added, 'I do not understand this squeamishness about the use of gas. I am strongly in favour of using poison gas against uncivilised tribes.'[14] General Sir Aylmer Haldane informed Churchill that poison gas would be more useful against the hilly Kurdish redoubts since in hot, open areas gas was more volatile.[15] In fact, the weapons used by the RAF in its civilising mission against the 'uncivilised tribes' were quite lethal. The British cabinet was reluctant to employ chemical warfare but Churchill convincingly argued the use of gas should not be prevented 'by the prejudices of those who do not think clearly'.[16] Eventually the go-ahead to use poison gas was given and an operation was mounted against the Iraqi rebels culminating in the collapse of the rebellion. Churchill later said the mission had an 'excellent moral effect'.[17]

The RAF was used to bomb the Kurds and Iraqis before, during and after the revolt. No remorse appears to have been shown by the British government for this action. As Wing-Commander Harris emphasised, 'The Arab and Kurds now know what real bombing means in casualties and damage. Within forty-five minutes a full-size village can be wiped out and one-third of its inhabitants killed by bomb or gases.'[18] It seems likely that Britain used the suppression of the Iraqi revolt in order to try out new chemical weapons. An Air Ministry list of available weapons for warfare in 1920 included, 'Phosphorous bombs, war rockets, metal crowsfeet [to maim livestock], man-killing shrapnel, liquid fire and delay action bombs'.[19] Certainly devices developed between 1918 and 1920 included the forerunners of napalm and phosphorous and fragmentation bombs.

Britain was not the only power to engage in the indiscriminate employment of chemical weapons. Spain and France also used poison gas in 1921, against the Berber rebellion in the Rif region of northern Morocco. The revolt, led by Abdel Krim, began in 1921 and beat the Spanish Army decisively in July of that year. In response the Spanish air Force took reprisals against the homelands of the tribes who had joined the rebels by making excessive use of poison gas. When the rebels continued to advance, proclaiming an independent republic of Rif, France dispatched 400,000 troops to aid the Spanish.[20] Entire Rif villages were wiped out by the French chemical weapon, aerial bombardment and artillery. The French Communist Party (PCF) apparently called them to stop 'immediately the spilling of blood'.[21] However, this is reformist writing of history suggesting the early communists were pacifists, which they were not. In fact, at the time, the PCF proclaimed its solidarity with the rebels and organised dock-workers' strikes, refusing to move war material to Morocco, and in October 1925 called a general strike against the 'colonial war'.

The movement for chemical arms control received its principal impetus from the development of chemical weaponry by all sides in the course of the First World War. However, the fact that the German Army had taken the first steps in the military deployment of poison gas was an important part of anti-German *Schrecklichkeit* (atrocity) propaganda. Certainly, at the end of the First World War the victorious Allies had decided to reaffirm in the Treaty of Versailles the pre-war prohibition of the use of poisonous gases and to forbid Germany to manufacture or import them. Undoubtedly, the continuing categorisation of such warfare as a German atrocity fuelled the very persistent diplomatic efforts to ban chemical warfare. Indeed, similar provisions were included in the peace treaties with Austria, Bulgaria and Hungary. Following such incidents as those detailed above, the Americans took the initiative to go a step further and attempted to introduce such provisions to apply on a more general scale. Drawing on the language of the treaties devised in Paris in 1918–1919, the Washington Disarmament Conference of 1922 introduced such a clause into a treaty on submarines. The US Senate gave its advice and consent to ratification without a dissenting vote. However, it never entered into force since French ratification was necessary and although the French were prepared to ratify the 'noxious gases' clause they objected to the submarine provisions.

At the 1925 Geneva Conference for the Supervision of the International Traffic in Arms and Ammunition, the United States once again took the initiative by seeking to include a clause prohibiting the export of gases for use in war. At France's suggestion it was then decided to draw up a protocol on the non-use of poisonous gases and, at the suggestion of

Poland, the prohibition was extended to bacteriological weapons. Such suggestions were, of course, a response to the use of poison gas during the First World War and the incidents in Iraq and Morocco in 1920 and 1921, and were then an attempt to prohibit further use of such chemical weapons. Indeed, the Geneva Protocol can be seen as the first diplomatic attempt at limiting chemical and biological warfare. However, the protocol concerned only the use of chemical weapons between states and so did not cover internal or civil conflicts. Indeed, even at the time many states had reservations about the wording of the protocol, especially regarding the right of retaliatory use for this made it effectively a no-first-use treaty. Nevertheless, the final document signed on 17 June 1925, and in force from 8 February 1928, recognised the significance of bringing together controls on chemical and biological weapons. However, although it prohibited the use of such weapons, it did not prohibit basic research, production or possession. Additionally, there was no verification mechanism contained within the protocol and compliance was voluntary. Thirty-eight countries prior to the Second World War, including all the great powers except the United States and Japan, ratified the Geneva Protocol.[22] When they ratified or acceded to the protocol, some nations, including the United Kingdom, France and the Soviet Union, declared that it would cease to be binding on them if their enemies, or the allies of their enemies, failed to respect the prohibitions of the protocol.

The protocol contained major loopholes, most noticeably, the lack of verification or enforcement clauses and the rather vague use of the term 'other gases'. Consequently the major powers continued to develop chemical weapons in secret. During the late 1920s, for example, the Soviet Union began to develop their own chemical warfare capability with co-operation from Weimar Germany and, in the same period, the Japanese obtained their own chemical warfare capability. Indeed, the Japanese were industrious in their chemical weapons efforts producing mustard gas, Lewisite, chemical bombs, rockets, anti-tank grenades using hydrogen cyanide charges and other weapons. These were to be fully tested against the Chinese a little more than a decade later.

In 1935, Italy, a signatory of the Geneva Protocol, notoriously used poison gas in its conquest and occupation of Abyssinia (now known as Ethiopia) in East Africa. On 3 October 1935, Benito Mussolini launched an invasion on Ethiopia from its neighbours – Eritrea, an Italian colony, and Italian Somaliland. Ethiopia immediately protested to the League of Nations, which in turn imposed limited sanctions against Italy. These sanctions, arguably, although not crippling, put pressure on Italy to either win the war or withdraw. Mussolini believed the initial Italian offensive had not been pursued with proper strength and, consequently, the Italian commander was replaced. The new Commander, Marshal Pietro Badoglio, in an effort to defeat the Ethiopian troops led by Emperor Haile Selaisse, quickly resorted to chemical weapons. Despite the Geneva Protocol, which Italy had ratified in 1928 and Ethiopia in 1935, the Italians both dropped mustard gas bombs and sprayed mustard gas from aeroplanes. They also used mustard gas in powdered form as a 'dusty agent' to burn the unprotected feet of the Ethiopians. Additionally, there were also rumours of phosgene and chloropicrin attacks, but these have never been verified. The Italians attempted to justify their use of chemical weapons citing the exception of the Geneva Protocol restrictions that referenced acceptable use for reprisal against illegal acts of war. Italy stated that the Ethiopians had tortured or killed Italian prisoners and wounded soldiers.[23] Chemical weapons were devastating against the unprepared and unprotected Ethiopians. With few anti-aircraft guns and no air force, the Italian aircraft ruled the skies. Haile Selaisse emotionally described the nightmare to the League of Nations:

> Special sprayers were installed on board aircraft so they could vaporise over vast areas of territory a fine, death dealing rain. Groups of nine, fifteen or eighteen aircraft followed one another so that the fog ensuing from them formed a continuous sheet... soldiers, women, children, cattle, rivers and lakes were drenched continually with this deadly rain... these fearful tactics succeeded. The deadly rain that fell made all of whom it touched fly shrieking with pain. In tens of thousands the victims of Italian mustard gas fell.[24]

This view is reinforced by Piers Brandon: 'They [the Ethiopians] had no war-planes with which to retaliate, hardly any artillery and little ground support... The Ethiopians had no coherent strategy and no fixed chain of command. All they had was common purpose and boundless courage.'[25]

In May 1936 the Italian Army completely routed the Ethiopian Army and, indeed, Italy controlled most of Ethiopia until 1941 when the British and other Allied troops re-conquered the country. Major Norman E. Fiske, an observer with the Italian Army, commented with regard to the role of chemical warfare that the Italians were clearly superior and victory for them was assured no matter what. According to Fiske, the use of chemical weapons was, however, nothing more than an experiment.

He concluded in his report: 'From my own observations and from talking with [Italian] junior officers and soldiers, I have concluded that gas was not used extensively in the African campaign and that its use had little effect on the outcome.'[26] Others, who felt the Ethiopians had made a serious mistake in abandoning guerrilla operations for a conventional war, supported Fiske's opinion. Captain John Meade, an observer with the Ethiopian Army, on the other hand, thought, contrary to Fiske, that chemical weapons were a significant factor in winning the war. According to Meade, chemical weapons destroyed the morale of the Ethiopian troops who had little or no protection. Additionally, chemical weapons were used to attempt, and successfully achieve, the break-up of troop concentrations. Captain Meade's report concluded: 'It is my opinion that out of all of the superior weapons possessed by the Italians, mustard gas was the most effective...it temporarily incapacitated very large numbers and so frightened the rest that Ethiopian resistance broke completely.'[27] Major General J.C. Fuller, assigned by the League of Nations to the Italian Army, highlighted the use of mustard gas to protect flanks of columns by denying ridgelines and other key areas to the Ethiopians. He concluded: 'In place of the laborious process of picketing the heights...those sprayed with gas were rendered un-occupiable to the enemy, save at the greatest risk. It was an exceedingly cunning use of this chemical.'[28] B.H. Liddell-Hart, another military expert, compromised between the two schools of thought and concluded: 'The facts of the campaign point unmistakably to the conclusion that mechanisation in the broad sense was the foundation on which the Italian's military superiority was built, while aircraft, the machine gun and mustard gas proved the decisive agents.'[29] One outspoken critic was the author Evelyn Waugh who believed Italy, as the 'Catholic Crusader', had far more right to control Ethiopia than the natives. Most controversial of all, when the Italians began dropping poison gas on Ethiopia, Waugh played down the effect: 'Gas was used but accounted for only sixteen lives.'[30] However, generally it seems that most observers agreed that the Italians would have eventually won, whether chemical weapons were used or not, and it has been contended that, in general, the British and Americans learned little from this war. Porton Down's Chairman's Annual Report in 1937 concluded that the use of gas in Ethiopia did not disclose any new chemical warfare tactics, but only confirmed existing tactical use expectations.[31]

Two main features characterise the manner in which gas seems to have been used in Ethiopia. First, aircraft-delivered chemical weapons were employed, and second, as far as the employment of mustard gas is concerned, the bomb quickly gave way to spray tank as the principle means of delivery. A possible conclusion from the first of these points is that, given the extended supply lines of its expeditionary army, the Italian High Command felt that while it was not worth giving its ground forces a chemical warfare capability, aero-chemical weapons were worth having for some sort of ground support role. From a technical point of view, the performance of the bomb does not appear to have been very great: it provided ground contamination within a radius of 10 or 20 m, but the bomb fragments were generally visible and there was little aerosol effect. Ethiopian troops learned to avoid crossing ground in the immediate vicinity of bomb fragments, so, unless the bombs fell on top of troop formations, they had little effect. Against mustard gas spray tanks, however, evasive action was much more difficult. If several aircraft were used to deliver the attack, the area covered by the spray was too great for people to escape contamination, unless they were at its edges, and there were no bomb fragments to warn people against crossing previously contaminated ground. All in all, the Italians appear to have put their mustard gas to three main uses: to protect the flanks of advancing columns, to disrupt Ethiopian communication centres and to demoralise Ethiopian troops. All three uses can certainly be seen as an extension of the mustard gas artillery shelling doctrine established by the Germans during the First World War.

Some writers believe that chemical warfare had a decisive effect on the outcome of the war; others hold it was merely a useful tactical aid, not of any major significance by itself in determining the outcome. The former group takes the view that Italy, harassed by League of Nations sanctions, was anxious to bring the war to an end as soon as possible. The Italian campaign had not progressed far enough by the time the rainy season was approaching and, as fighting would have then become impossible, extraordinary measures had to be taken. Gas was therefore used to demoralise the unprotected Ethiopians, and to break their resistance once and for all. Whether or not the rains would have immobilised the Italian Army, this view probably exaggerates the power of the League of Nations sanctions. Possibly if gas had not been used, the Italians might not have reached Addis Ababa in one campaign, but they would probably have got there sooner or later unless outside powers took more drastic steps to stop them. In addition, chemical warfare techniques were not introduced as suddenly as this explanation would require; rather they were used to accomplish a gradually increasing range of tactical objectives in the face of unexpectedly stiff resistance. In this way the war was probably shortened but its outcome was not seriously affected.

There was no scarcity of reports of chemical warfare during the Spanish Civil War, for the recently concluded Italy–Ethiopian War had demonstrated vividly the ability of such allegations, whether founded in truth or not, to stir public feelings. There were reports of shipments of gas from Hamburg, and counter-reports of shipments from Black Sea ports.[32] Emergency appeals for supplies of gas masks were launched on several occasions by both right-wing and left-wing organisations throughout Europe. Nevertheless, the only report which had any ring of truth about it came during the early stages of the war, when a well-regarded London newspaper quoted both an insurgent spokesman and an 'observer on the government side' on an incident in August 1936. Thirty-four guns of government artillery, it was reported, fired tear-gas shells against insurgent positions on the Guadarrama front.[33] Subsequently, newspapers reported threats by the insurgents to retaliate with their own stocks of gas. The inhabitants of Madrid were stated to be expecting their city to be gas-bombed[34] and, indeed, reports in December stated that Madrid had been shelled with chemical warfare casualty agents, and that in the previous month government forces had used gas in a massive attack.[35] None of these reports were ever validated.

The next war that drew the interest of chemical warfare experts was the Japanese invasion of China in 1937. The Japanese, in addition to their biological work, had an extensive chemical weapons programme and were producing agents and munitions in large quantities by the late 1930s. During the war with China, Japanese forces used chemical weapons in the form of shells, tear-gas grenades and lactrimatory candles, often mixed with arsenic smoke screens. By 1939 the Japanese had reportedly escalated to the use of mustard gas and Lewisite. Against the untrained and unequipped Chinese troops, the weapons proved effective. The Chinese reported that their troops retreated whenever the Japanese used just smoke, always fearing it was a chemical attack.[36] A Soviet authority summarised the part played by gas in the Sino–Japanese War thus: 'Japanese units active in China included twenty-five per cent chemical projectiles in the complement of artillery forces, and in the store of aviation munitions thirty per cent were chemical bombs. In several battles up to ten per cent of the total losses suffered by the Chinese armies were due to chemical weapons.'[37] A post-war survey of the reports and allegations of chemical warfare referred to the use of CN, DA, DC, phosgene, diphosgene, chloropicrin, hydrogen cyanide, mustard gas and Lewisite. The weapons said to have been used included aircraft bombs, artillery shell and toxic candles.[38] However, Japanese officers under interrogation at the end of the Second World War denied that chemical casualty agents had ever been used, but then the imminence of the Japanese War Crimes Tribunal would hardly have encouraged veracity on this point, if such agents had been used. The use of irritant agents, however, was freely admitted, but the Japanese did not regard them as being prohibited by international law because they caused neither death nor permanent injury.[39]

The principle advance in chemical warfare theory made by the Italians in Ethiopia can be seen to be the demonstration with aircraft spray tanks that mustard gas could be used to military advantage. The Japanese experience in China supported this, and in addition, showed how irritant agents could be effectively used by ground forces in mobile warfare situations, provided the enemy was unprotected. Certainly, the Japanese Army Chemical Warfare Establishment viewed the China campaign as a valuable opportunity both to verify its assessment of the possible uses of chemical weapons and to secure their wider acceptance throughout the army. The Japanese chemical warfare techniques had little originality over those used during the First World War and the Italy–Ethiopian War. However, there was one technique that they appear to have used frequently and which departed from existing practice. This was the use of irritants for purposes other than the simple harassment of the enemy; in many respects, some of the employment techniques seem comparable to those later used by US forces in Vietnam.

However, the Japan–China conflict seems to have gone almost unnoticed in the Western press, and it was after the Italy–Ethiopian War that the possibility of a chemical war in Europe became the primary concern of both the British and US Army's Chemical Warfare Services. Consequently, they studied the chemical warfare capabilities of both Germany and Italy, but, as will be seen, intelligence completely missed the German development of nerve agents.

Possibly the earliest recorded use of a substance that worked like a nerve agent, by inhibiting cholinesterase, is by native tribesmen of western Africa who used the Calibar bean as an 'ordeal poison' in witchcraft trials. An extract, the elixir of calibar, was later used medicinally, and in 1864 the active agent was isolated and called, first 'physostigmine' and later 'eserine'. During the next 80 years, chemists made numerous advances in organophosphorus chemistry but generally they did not realise the toxicity of the substances with which they were working. It was not until the end of 1936 that Dr Gerhard Schrader, a German scientist at the huge German chemical giant IG Farben, who was researching the use of organophosphorus compounds as insecticides, accidentally stumbled on a series of poisons which, when mixed, had phenomenal intensity. By December he was ready to test the chemical for the first time; he did this by using one part in 200,000 on leaf lice. All the insects were killed. In January 1937 the first manufacturing trials began, but almost immediately he realised that the new, promising insecticide had side effects which were 'extremely unpleasant'.[40] Schrader and his assistant were forced to stop work for three weeks after suffering symptoms such as diminished sight and breathlessness. In fact, they were lucky to escape with their lives because inadvertently they had discovered the world's most powerful weapon, the original nerve gas 'tabun'.

In tests that spring, almost all the animals exposed to even tiny quantities of the new 'insecticide' were dead in twenty minutes. It now became obvious that Schrader's discovery could not be used for its original purpose. Instead, under a Nazi decree of 1935 requiring German industry to keep secret any invention with military potential, Schrader was summoned to Berlin to demonstrate tabun to the Wehrmacht.[41]

Tabun's value as a chemical weapon was quickly recognised, and the patent applications covering it were made secret.[42] Tests conducted on dogs or monkeys who had been exposed to tabun showed they lost all their muscular control, their pupils dilated, they frothed at the mouth, vomited, had diarrhoea and, finally, within 10 minutes they went into convulsions and died.[43] In addition to its obvious potency, tabun had other advantages: it was colourless, virtually odourless and could poison the body by inhalation or skin penetration. This nerve gas was a great advance over the chemical weapons of the First World War. IG Farben, who realised the moneymaking potential of Schrader's new discovery and in order to allow him to pursue his research into organic compounds undisturbed, moved him to a new factory at Elberfeld.[44] A year later, in 1938, he discovered sarin which, when tested on animals, was found to be ten times as poisonous as tabun.[45] In 1944, during the course of work for the army on the pharmacology of tabun and sarin, Dr Richard Kuhn, the Nobel Laureate, prepared the pinacolyl analogue of sarin, and this substance, 1,2,2-trimethylpropylmethylphosphonofluoridate (soman), was found to exceed sarin in toxicity. By this stage in the war, however, it was too late to complete the necessary development work on the new agent and, in any case, the pinacol needed for its manufacture was in short supply.[46]

The Germans succeeded in concealing their work on organophosphorous compounds throughout the inter-war and war periods, and although in British and American laboratories a variety of such substances were synthesised, the nerve gases themselves were not found. The nerve gases, called 'G-agents' after the marking on the Second World War weapons containing tabun, were the only really significant advance in chemical warfare since the First World War.

The roles of the British and American chemical industries in the inter-war years were, overall, relatively limited ones until the threat of another war appeared in the mid-1930s. Then, as in the First World War, many calls were made upon the industry as the nations prepared to defend themselves, for Britain from 1939 and for the United States from 1941.

In Britain, the Services and Chemical Warfare Committee had laid down the general programme of research and development required in 1920. The main areas were individual respiratory protection, the collective protection of His Majesty's Navy, the design of more efficient weapons and munitions, especially aerial gas bombs, the meteorology of gas and smoke clouds, and the treatment of gas casualties. The annual reports from 1921 to 1938 provide a detailed account of the progress of the programme at Porton, and Sutton Oak. They should also provide the definitive account of UK policy, doctrine and planning up to the Second World War,[47] but regrettably these documents are incomplete, some still retain restricted access, and other files of interest were destroyed when the major military presence left Porton Down in 1957.

One of the major post-war tasks had been the assessment of the condition of over a million respirators. One per cent of the stock was sent to Porton for penetration tests; however, the only convenient method was to do this by having men breathe through them whilst exposed to the irritant arsenical smoke DM (diphenylamine chloroarsine). These tests imposed some physical and psychological strain on the staff because the results were urgently needed and because the respirators were found to be largely penetrable and useless through deterioration. The deficiencies triggered requests for improved designs of respirators, and by 1926 the common service GS respirator became the standard pattern for the armed forces until the 'Light Type' respirator supplanted it in 1942. Another important feature of post-war research was that into protection of the hands, body and feet from mustard gas droplets. During the First World War, the respirator alone had largely sufficed as the soldiers soon learnt to keep away from squashed liquid mustard gas. In the First World War there was no prospect of a more insidious exposure to small droplets of mustard gas sprayed from aircraft. Mustard gas vapour was effectively kept out of the eyes and respiratory tract by the respirator. However, as the air forces of the major nations developed and the feasibility of aerial bombing attacks on cities grew, Air Raid

Precaution became an increasing national concern. The Italian use of mustard gas was a further harbinger of concerns that steadily increased, not least because Italy was a signatory to the Geneva Protocol.

Other products of Porton Down's research and development at the end of the 1930s included eye shields to protect against high-altitude liquid mustard gas attack, the oil-skin anti-gas cape, impregnated battle dress, protective 'dubbing' for boots, detectors and detector paints, decontamination procedures, gas identification sets for service units, respirators and anti-gas covers for horses and dogs. Protection for camels was also studied; a prototype respirator still exists in the establishment at Porton Down.

There were probably no better-equipped forces in respect of anti-gas defence than those of the United Kingdom in the late 1930s. Britain had emerged from the First World War with a primitive respirator and basic techniques for gas-proofing dugouts, and little else. At the end of the 1930s, superior-quality anti-gas equipment was available to the armed forces to cater for all known hazards and a cheap, but efficient, respirator had also been developed for the civilian population.[48] However, as far as offensive capabilities were concerned, investment had been limited and production had been minimal in terms of agents and weapons due to political unease and uncertainties. By 1938 the international situation was such that offensive research and development and the production of war reserve stocks of mustard gas were authorised by the British Cabinet, albeit that it was realised that a useful production output could not be obtained for at least 12–18 months.

Comparatively many historians have concluded that the US Army on the eve of the Second World War was totally disorganised. While this might be true in a broad sense, parts of the army had begun to get ready for war well before the Japanese attack on Pearl Harbor in 1941 plunged the United States into conflict. That was certainly the case for the Chemical Warfare Service and, indeed, its pre-war preparations, like those at Porton Down, proved invaluable in enabling it to meet the challenges of total war. It was the National Defence Act of 1920 and the Procurement Act of 1921 that allowed the storage and distribution of war materials, including chemical items. Indeed, on 9 December 1921 the Assistant Secretary for War approved a list for the Chemical Warfare Service, which included toxic agents and smoke and cloud gas materials. However, the supplies then to hand would not have gone far, and so in the mid-1930s when Europe began to show distinct signs of approaching war, action had to be taken. During the 1930s the Chemical Warfare Service had developed a stockpile which it was realised was inadequate.

More money was needed but this was only made available after the German invasion of Poland in September 1939. The main problem in the pre-war planning period was that those entrusted with the planning had made some incorrect assumptions. They assumed that mobilisation would be slow and orderly, apparently forgetting the chaotic mobilisation of the First World War.

The heart of the American chemical offensive capability was the chemical mortar. However, in December 1941 there were only 44 chemical mortars available, but this was quickly rectified as the demand for this versatile weapon increased. By 1943 the M2 series 4.2-in. chemical mortar had been developed and this rapidly became the central weapon of chemical warfare units, not only for chemical delivery if needed, but also for high explosive, smoke and white phosphorous rounds.[49] The United States Army Air Force (USAAF) had 100-lb mustard agent bombs, 500-lb phosgene or cyanogen chloride bombs and 1000-lb phosgene or hydrochloric acid bombs. An important point, frequently forgotten, was that it was necessary to take along the full spectrum of chemical weaponry wherever American troops were deployed; ultimately the positioning of chemical weapons in forward areas, in case of need, would result in one major and several near disasters.

At the beginning of the war American troops were issued with the M1 or M2 training respirator. However, these masks with their moulded rubber facepieces and heavy canisters caused two significant problems for the military. First, the shortage of rubber during the Second World War meant a substitute had to be found. Secondly, the weight of the mask with the canister had to be reduced, particularly for amphibious assaults. By 1944, with a major invasion of Europe pending, the army requested a better assault mask. To meet this requirement the Chemical Warfare Service turned to the original German First World War mask that put the canister directly on the facepiece. The result was the M5 Combat Mask which was standardised in May 1944. Due to the shortage of rubber, the M5 mask was the first to use synthetic rubber (neoprene) for the facepiece. The canister used ASC Whetlerite charcoal, which provided better protection against hydrocyanic acid, a chemical agent found in a Japanese grenade shortly after the attack on Pearl Harbor.[50] As in Britain, a major civil defence programme to protect civilians was also initiated in the United States. Of particular concern were protective devices for children and with the help of Walt Disney, a Mickey Mouse gas mask was designed in the hope that children would not be frightened if they had to wear it.[51]

Other products of the American Chemical Warfare Service's research included a Vapour Detector Kit and, indeed, the refined model of 1943 proved to be one of the most significant developments of the Chemical Warfare Service during the war, described in news releases as being as 'effective as a modern burglar alarm'.[52] Additionally a sprayer decontamination apparatus was developed and for the treatment of gas casualties a protective ointment kit, which included the ointment and a tube of British Anti-Lewisite (BAL) eye ointment, was mass produced, 25 million in all.

And so, as the probability of another war increased, all the major powers rapidly expanded their efforts in the study of chemical warfare. These efforts embraced every aspect of the subject from the search for new and more deadly chemical agents to research and development directed towards the provision of even better defensive measures to protect against known agents. One much discussed topic was the role that aircraft would take in the next chemical war. It has been said that during the First World War the air forces of all the belligerents refused to participate in the gas war, although the Chiefs of Staff recognised the utility of air-delivered chemical weapons. A certain amount of development work on such weapons was carried out; for example, in early 1915 the British were experimenting with bombs charged with hydrogen cyanide and the commander of the BEF called for chemical bombs as part of his retaliatory chemical warfare material.[53] The Allied and Central Powers accused each other of using aerial gas bombs during the war, but the reports of such incidents were either unconfirmed or stated to be false.[54] Civil Defence authorities in Paris and London had made preparations to meet a chemical attack. In June 1915 there were plans to issue service respirators to police officers in Paris against possible Zeppelin gas raids.[55] The British War Cabinet set up a committee to study the air defences of the United Kingdom after the heavy attacks on London in July 1917, believing gas bombs would probably be used in subsequent raids.[56] The fear proved unfounded.

There is, however, some evidence that the strategic gas bombing of cities was being contemplated during the First World War. Professor Fritz Haber, speaking in Berlin in 1926, spoke of suggestions for the gas bombing of Verdun. The Chief of the German General Staff, Haber turned this down because the techniques of Zeppelin bombing were too inaccurate at the time.[57] Another authority refers to plans made by the British Independent Bombing force during the closing stages of the war to include gas in the bomb loads for an attack on Berlin.[58] However, there is no indication that a decision was ever made to initiate strategic gas bombing.[59] Apart from the dislike of chemical warfare shown by the First World War air forces, a strong explanation for the absence of aerial gas warfare is simply that the available weapons were not good enough.

Post-First World War exploration of aircraft-delivered chemical weapons saw aircraft payloads increased, so that if chemical weapons were to be used from the air they could be employed against large targets. The chemical arm of the air force, therefore, attracted the attention of popular writers, both serious and scaremongering. The advent of the strategic bomber was seen as the forerunner of total warfare in which civilians and military alike would be targets of attack. With bomb loads of rumoured new-found chemical warfare agents, the mass destruction of whole continents was predicted.[60] Official attitudes seemed to confirm this nightmare. In the United States it was announced that 200 tons of phosgene dropped in bombs would be enough to kill every occupant of an area 100-miles square.[61] One aeroplane carrying 2 tons of a newly discovered percutaneous agent, Lewisite presumably, could spray an area 100-ft wide and 7-miles long, with enough agent to kill every man in it.[62] Fries predicted: 'The dropping of gas bombs of all kinds upon assembly points, concentration camps, rest areas . . . will be so fruitful a field for casualties and for wearing down the morale of armies in the future that it will certainly be done, and done on the very first stroke of war.'[63] Indeed, as a test of the use of aircraft in a chemical war, the first simulated attacks against battleships took place as early as 1921.[64]

During the 1920s and 1930s then, many people felt that the rise of aero-chemical warfare had provided the means for destroying life on an unprecedented scale. Certainly some of the predictions made at the time credited chemical weapons with mass-destruction capabilities comparable to those made nowadays. Typical predictions included estimates that 12 Lewisite bombs would eliminate the entire population of Berlin,[65] or that a single bomb dropped on Piccadilly Circus in London would kill everyone from Regents Park to the River Thames.[66] These assessments, however, did not coincide with those made by the military establishments, who generally rated the casualty-producing ability of chemical warfare agents as the same as that of conventional weapons. Indeed, the military believed that if they were forced to mount a chemical warfare agent attack, as a retaliatory measure, it would be a gesture rather than in the expectation of securing a strategic advantage.[67] Certainly, it was believed that there was little to be gained by killing civilians: it would be far more advantageous to destroy their homes, factories and transportation. However, the German discovery of nerve gases indicated the possibility of huge civilian gas casualties, both intended and unintended. Certainly, the British civilian gas mask of 1939, unchanged from the military model in use at the end of the First World War, would have provided no protection against the German nerve agents. Even if the Germans chose not to launch nerve gas attacks on civilian targets, the use of nerve agents against combatants would have inevitably killed great numbers of non-combatants as well. The changing role of the aeroplane in this period meant that as well as the methods used to disseminate chemical agents during the First World War, dispersion from the air by spray or bomb now became a possibility. Had chemical warfare been employed in 1939 or after, a chemical war in Europe could well have seen casualties in tens of millions.[68]

4
The Second World War

At the beginning of the Second World War, the experience of the First World War gave most of the combatants the expectation that chemical warfare would be used to an even greater extent, despite the Geneva Protocol having been put into place in the inter-war period. Chemical warfare had resurfaced in 1937 in Abysinnia (Ethiopia) and in China, and it was clear that another great conflict was nearing and chemical weapons were expected to be used, both on the battlefield and against civilian populations. Certainly air raids and gas attacks were expected as soon as war broke out and the British government issued the entire civil population, including babies, with gas masks – some 30 million in total. Perhaps in response to the D-Day landings in June 1944, it was not until late autumn 1944 that Hitler intervened in the matter of gas masks and appointed a special commissioner directly responsible to him. With great haste a programme was set up to protect the entire German population from the effects of gas warfare. Although gas mask production rose to more than 2,300,000 per month, it was evident it would take a while before the entire urban population would be properly equipped.[1] However, in Germany, since 1931 the Ministry of the Interior had been issuing guidelines for civil defence, and in 1932 the first release of the *Vorläufige Ortsanweisung für den Luftschutz der Zivilbevölkerung* was issued, which by the end of the war comprised 12 chapters with numerous comprehensive attachments. Similarly in Britain, from the mid-1930s, His Majesty's Stationary Office (HMSO) published *Air Raid Precautions* handbooks which, by the end of the war, comprised nine volumes. In addition, local authorities distributed a series of Air Raid Precaution (ARP) memoranda dealing with various aspects of air raid organisation. Newspaper articles and popular fiction predicted that chemical warfare would turn entire regions of Europe into lifeless wastelands.

However, despite the fact that by 1936 nerve gases had been developed in Germany, which were capable of killing on a previously unimagined scale, to everyone's surprise it did not happen. Why? Certainly gas, as had been demonstrated in the First World War, was fundamentally a siege weapon, intended to root out troops dug into trenches and fortifications. As the German *Blitzkrieg* was a war of rapid mobility it meant that gas attacks could hamper the attacker as much as it hurt the defender. As the Allies closed in, from east and west, and Germany's position became increasingly desperate, the pressure on the Germans to use anything they could to fight back increased tremendously, but even under these conditions they did not use gas on the Allies. Indeed, use of gas might have gained the Germans short-term advantage, but the overwhelming retaliation Hitler expected would, in his opinion, only have accelerated defeat. In fact, ultimately no major use of chemical weapons occurred, although rumours of incidents kept the chemical warfare services of all belligerents in a constant state of heightened alert. The possibility that a massive chemical attack could happen any day was reflected in media reports of the time. 'Military authorities have predicted gas will be used in the present war, if at any time the user can be sure of an immediate and all-out success from which there could be no retaliation.'[2] The technological advances made in the Second World War in the development of chemical agents are valid for further examination, if only because terrorist organisations in the world appear quite willing to deploy chemical weapons on today's battlefield, unlike their predecessors between 1939 and 1945.

The chemical industry was one in which it was impossible to separate warlike purposes from peaceful purposes. Just by continuing its normal activities of supplying materials to other industries the chemical industry made an indispensable contribution to war production. Among the more important direct applications were ammonia synthesis, which provided the basis for nitric acid for explosives, and dyestuffs technology which could be applied to war gases. However, many products that were essential in the peacetime industry, such as chlorine and cyanide, could also easily become chemical weapons. In fact, in general, it could be said that almost any branch of the chemical industry could be converted from peaceful to warlike applications; some certainly were from 1939, others, such as the production of specialised war gases, had no peaceful application at all.

When war broke out it was poison gas that captured the government and public imagination as one of the worst of the prospective horrors of war. Aside from the issue of gas masks, government responses to the possibility of gas attacks were not purely passive. Certainly the British government believed that if the Germans used gas then so should they; it seems, then, that no one had very much faith in the various declarations that had been made against its use since the First World War. Gas, therefore, was made in Britain by ICI and the first batch of KSK (tear gas) was, in fact, produced for the War Office as early as 1926.[3] However, production was accelerated between 1939 and 1940 and in this one year 30 tons of KSK, 145 tons of brombenzl cyanide (BBC), 650 tons of phenylarsenic acid (B.001) and 500 tons of thiodiglycol for mustard gas were manufactured.[4] Plans for much larger production quantities, especially of mustard gas, were also put in hand and although a new plant was made ready, it was never put to use.[5]

No doubt the British government had some scruples about the use of gas, and certainly by 1940 was well aware that in fact high explosive was a much more effective killer than any of the gases available to them. This was not the case in Germany. Starting in 1940 two new works were built, at Dyhernfurth and Gendorf, to exclusively produce mustard gas and the new lethal nerve agents, tabun and sarin.[6]

Many different types of chemicals were made and tested as possible chemical warfare agents in most advanced countries between 1939 and 1945 but all the information available suggests that the conclusions of Sartori, in his review of this subject, are correct and that the most important areas of chemical research were connected with organophosphorous compounds – nerve gases.[7]

In September 1939, as scientists prepared the first samples of sarin, the German Army launched its invasion of Poland. For the second time in a generation, German chemists were at the forefront of their nation's war effort. On 19 September Hitler told an audience in Danzig:

> Well, if England wants war she can have it. It will not be an easy war, as they like to think, nor a war fought in the way the last one was. It will be a war of such destructiveness as no one has imagined. We possess fearsome new weapons against which England will be defenceless...[8]

It is possible that Hitler had in mind the new nerve gas. After all, in the same month the German chemical industry was ordered to put in hand plans to build the new factory at Dyhernfurth which was hoped to be capable of producing 1000 tons of tabun a month.[9] A new company, Anorgana, was formed under the auspices of Otto Ambros of IG Farben,[10] and through Anorgana, Ambros provided the chemists and technicians needed to build and run the Nazi war gas plants. However, the Nazi nerve gas project was plagued by difficulties and it was not until 1942 that the Dyhernfurth works became operational. Even before production got underway there were over 300 accidents and at least 10 men were killed. If the Germans had any doubt about the potency of their nerve gases the Dyhernfurth accidents must have dispelled them. After all, if this was the effect in a laboratory what might the effects prove to be if the gas was unleashed on the battlefield against unprotected and unsuspecting soldiers?

During this period research on organophosphorus compounds had been proceeding in Britain on similar lines although no compounds as potent as tabun and sarin were discovered. Britain did, however, produce two substances that were sufficiently potent to merit detailed investigation: DFP and Fluoro-tabun. However, in 1945 research was still 'continuing'.[11] To illustrate the importance of the discovery of the German nerve gases for chemical warfare, their toxicities are set out in Table 4.1 in comparison with British discoveries of the same period and those of the earlier chemical warfare agents.

There is no doubt that by the middle of the war the Nazis had acquired vast, hidden armouries of chemical weapons and the Wehrmacht still found millions of marks to pump into the testing and production of poison gas. Indeed, the effort put by the Germans into chemical warfare research was considerable with them employing double the number of scientists than Britain,[12] and their twenty factories were capable of producing about 12,000 tons of poison gas a month.[13] Indeed, the Allies believed, in a report issued after the war, that the Germans had about 70,000 tons of poison gas stockpiled at various

Table 4.1   Comparison of toxicity of nerve gases and other agents

Agent	Median lethal dose (mg)[a]	Lethal dose (mg)
Tabun (German)	100–200	200–400
Sarin (German)	50–100	100–200
DPF (British)	1500–2000	–
Phosgene	3300	–
Mustard gas	1500	4000–5000

[a] Median lethal is the dose which would be expected to kill 50 per cent of the population subjected to it.

Note: Compiled from Karl Heinz Lohs, Synthetic Poisons: Chemistry, Effects and Military Significance, 2nd edn, East German Military Publishing House (1963). Published in translation by United States Department of Commerce, OTS as JPRS, 23.681 (1964), p. 21.

locations around Germany.[14] In April 1945 the British captured a German ammunition dump that contained 105-mm shells marked with a single green ring and the legend 'GA'; they were filled with tabun. Other dumps were found with a total of half-million shells and 100,000 aerial bombs filled with nerve gas.[15] The Allies, however, should not have been surprised to discover the nerve gas since throughout the war unsubstantiated rumours had circulated of a new German poison gas. Indeed, on 11 May 1943 a captured German proved to be a chemist from the main Nazi chemical warfare laboratory at Spandau. He told the British everything he knew of a super gas called 'Trilon 83' (tabun).[16] The informant told of a 'clear colourless liquid with little smell... which cannot be classed with any of the other war gases as it is a nerve poison'.[17] All the information passed on by the chemist was produced in a report that was circulated throughout Whitehall and Porton Down. Nothing happened. Paxman and Harris discovered the yellowing M1 19 report in a pile of de-classified government documents entitled 'Chemical Warfare Intelligence 1919–1944' in the late 1970s. Certainly they make it clear that the British were 'reliably' informed of the existence of nerve gas two years before the end of the war.[18] If Hitler had decided to use tabun or sarin in 1944, the decision to disregard the report would, undoubtedly, have gone down as one of the costliest oversights of the Second World War.

In addition to tabun, and sarin, the Germans had developed two types of mustard gas (*Lost*) – for hot climates *somer-lost* and for cold climates *winter-lost* – and an incendiary gas, N-Stoff (chlorine trifluoride), which was produced exclusively for the SS.[19] N-Stoff could cause clothes, hair and even asphalt to burst into flames. In 1923 a new disinfectant based on hydrochloric acid had been placed on the market in Germany, under the trade name Zyklon B by the *Deutsche Gesellschaft für Schädlingsbekämpfung* (German Pest Control Corporation) known more commonly as Degesh.[20] In the summer of 1941, Heinrich Himmler gave Rudolf Höss the task of organising the extermination of the Jews sent to Auschwitz. Höss made enquiries about methods employed elsewhere, particularly those for killing Jews and Gypsies by using exhaust fumes from motorised vehicle engines. Höss felt this procedure was too slow and not sufficiently reliable. However, during his absence from Auschwitz on his fact-finding mission, his deputy Karl Fritzsch tried to kill a number of Russian and Polish prisoners by other means – Zyklon B. When Höss returned he was shown that if the prisoners were shut in an improvised gas chamber and a certain amount of Zyklon B gas pellets were introduced, death was fairly quick. The main characteristic of

Zyklon B was the ease with which it penetrated the mucous membranes of the mouth, nose, oesophagus, stomach and lungs and then entered the bloodstream. Certainly Zyklon B was 6 times more toxic than chlorine, 34 times more toxic than carbon monoxide and 750 times more toxic than chloroform. Additionally, with each inhalation the person retained more of the gas and so the concentration of poison in the air was reduced. This fact simplified ventilation problems afterwards. Höss judged this result satisfactory and, after this, the use of Zyklon B was adopted as the method of mass murder at Auschwitz. At the Nuremberg Trials, after the war, much was made of the bills of lading for deliveries of this fumigant. It was argued that most of the concentration camps used Zyklon B to kill vermin and fight disease, thereby reducing the camp death rates. Other camps, designated by the experts at Nuremberg as extermination camps, used Zyklon B to kill inmates and thereby increase death rates. On this reasoning, deliveries of Zyklon B to Dachau were beneficial, while deliveries of Zyklon B to Auschwitz were criminal. The nerve gases, tabun and sarin, were almost certainly tested on the inmates of the concentration camps, although, as is to be expected, thousands of files on chemical warfare were destroyed between 1944 and 1945 so there is no official record of such tests.[21] However, it appears unlikely that the Nazi leadership would have agreed to the diversion of such huge sums of money from the economy for the development of a chemical warfare agent if it had not been shown overtly to be capable of killing men.

Additionally, the Germans developed a series of quite ingenious devices to deploy these new weapons. To slow up an enemy advance, for example, mustard gas was to be poured into holes in the ground, covered over and once the enemy broke the crust the mustard would be released with the obvious consequences.[22] A machine gun was tested capable of firing 2000 rounds a minute of ammunition filled with tabun or sarin, and the *Luftwaffe* had half-a-million 'gas bombs' including 750-kg phosgene bombs.[23] As early as 1939 the Commander in Chief of German chemical troops had advocated the use of gas against industrial concentrations and large cities as a weapon of terror.[24] Indeed there can be no doubt that had such an attack been undertaken, London, for example, would have been plunged into turmoil and enormous pressure would have been brought to bear on Churchill's wartime government.

The Germans also carried out a series of tests charging their V1 and V2 rockets with poison gas. Now then, with the V-weapons, German chemists had the means to deliver the terror that Hitler desired. Indeed, preparations were in hand to fill the V1 with phosgene, and tentative plans were put forward to deliver tabun in V-weapons right into the heart of London.[25] The British standard civilian respirator, not much improved on the military issue of 1918, might have offered some protection against phosgene but would certainly have offered no protection against tabun. In 1944, Hitler was sending waves of V1 rockets across the English coast at the rate of 200 per time, and had they been filled with tabun, Hitler could have had a weapon of horrifying magnitude. It is unclear if such a weapon could have been put to effective use. Tabun, of course, is hideously toxic, but it is also volatile and large concentrations of it are needed to be really effective. Such concentrations would have been difficult to achieve with the V1 rocket. Mustard gas, being persistent, would have been a bigger nuisance, but high explosive, as used in 1944, was probably more destructive, intimidating and effective.

By the middle of 1943 as the German war machine was on the decline, following, in particular, the defeat at Stalingrad, why then did Hitler not deploy his *Siegwaffe* (victory weapon)? By the middle of 1944, Germany had a formidable nerve gas arsenal dotted around the country containing enough tabun to kill the entire population of London as well as large stockpiles of more traditional chemical agents. So, surely as greater and greater tonnages of the weapons were stockpiled, the temptation to use them must have been correspondingly increased. But still Hitler failed to give the order to deploy them. Why?

Hitler, having himself suffered the effects of a British mustard gas attack in the First World War, was known to have a particular aversion to using chemical weapons. Indeed, from the available documentation it appears he only once visited a chemical warfare establishment in 1942. Nevertheless, as Germany's plight became more and more desperate during 1944 there is evidence that Hitler hoped that the nerve gases, if deployed, might still turn the war in his favour. Indeed, shortly before D-Day (6 June 1944) he bragged to Mussolini of secret weapons that could turn 'London into a garden of ruins' and he referred specifically to a deadly new war gas.[26] There is no doubt that by 1945 it would have been suicidal for Hitler to deploy his chemical weapons, even if he had had enough bombers left to launch the attack. Nevertheless, certainly Bormann, Goebbels and Ley repeatedly urged Hitler to deploy the nerve gases. Speer, the Minister of Armaments, on the other hand, remained silent. Later he stated: 'Hitler...had always rejected gas warfare; but now he hinted...that the use of gas might stop the Soviet troops...when no-one spoke up in agreement, Hitler did not return to the subject. Undoubtedly the Generals feared the unpredictable consequences.'[27]

Certainly Speer, believing that should such an attack be launched in 1945 it would cause the Allies to retaliate in kind, had been going to great lengths to divert raw materials away from the chemical warfare factories.[28] As he later testified at the Nuremberg War Crimes Trials in 1946:

> I was not able to make out from my own direct observations whether gas warfare was to be started but I knew ... they were discussing the question of using our two new combat gases: tabun and sarin ... They believed that these gases would be of particular efficacy and they did in fact produce the most frightful results ... For the manufacture of this gas we had three factories, all of which were undamaged and which in November 1944 were working at full speed. When rumours reached us that gas might be used, I stopped its production ... All sensible army people turned gas warfare down as being utterly insane, since, in view of their [the Allies] superiority in the air, it would not be long before it would bring the most terrible catastrophe upon German cities.[29]

However, had Hitler launched such an attack at the end of 1943, things might have been very different. The Germans, like the British, were aware that amphibious landing forces presented particularly vulnerable targets to gas attack during disembarkation and while struggling to establish beachhead positions. Clouds of non-persistent agents could cause enormous casualties among the massed troop concentrations as they moved up the beaches, while persistent agents could greatly complicate the intricate supply arrangements needed to support the landing. In the case of the D-Day landings, in June 1944, there were reports that the Germans had issued orders to use gas, but that for a variety of reasons the orders were countermanded.[30] Indeed, the Allies were so certain that the Germans had no war gases that during the D-Day landings all anti-gas equipment was left behind in Britain. Had Hitler used tabun against the bridgeheads in 1944 it may well have stopped the landings in their tracks. As General Omar Bradley later wrote: 'When D-Day ended, without a whiff of gas, I was relieved. For even a light sprinkling of persistent gas on Omaha Beach would have forced a decision in one of history's climatic battles.'[31]

The Chief of the US Army Chemical Warfare Corps, writing in 1946, expressed the view that heavy gas attacks on the Allied beachheads might have delayed the invasion of Europe for up to six months and, indeed, have made landings elsewhere a necessity.[32] Certainly, had the invasion been delayed the Germans would have had the time to complete the new V-weapons, which would have made the Allies' task harder and England's long-range bombardment considerably worse.

Certainly, the incentives for the Germans to use gas to repel the Allied landings in France were considerable. However, the principal restraint was fear of Allied retaliation in kind against German cities, and in this case the restraint was far more compelling than it had been against possible British use of gas to repel the threatened German invasion of the United Kingdom in 1940. German leaders must also have realised that if they initiated chemical warfare, Allied retaliation might not be confined to German targets: the United States would be given justification to use gas against Japanese targets. Behind both these constraints, of course, there also remained the uncertain influence of Hitler's personal attitude towards gas.

So it seems that had the chemical weapons in Hitler's arsenal been deployed, even as late as June 1944, it may well have seriously affected Britain's commitment to the war. Certainly Stalin, already furious that the Second Front had not been launched two years earlier, may well have sought a negotiated peace with the Germans. Indeed, tabun may well have saved Germany from defeat. There is no doubt, however, that by 1945 it would have been suicidal for Hitler to deploy his chemical weapons, even if he had had enough bombers to launch such an attack. So, although the incentives to use gas during the fighting in Europe perhaps reached their highest point when the Germans were trying to counter the Allied landings in 1944, the constraints were still more compelling.

Therefore, the reason Hitler failed to deploy his 'victory weapon' is quite simple. In 1943, Otto Ambros, of IG Farben, told Hitler that he had no doubt the Allies also had chemical weapons and, additionally, they could out-produce Germany.[33] Hitler pointed out that tabun was unique, but Ambros responded by suggesting that it was also 'known abroad'.[34] Ambros elucidated on this point by explaining that the properties of tabun and sarin had been revealed in scientific journals before the outbreak of war, indeed, possibly as early as 1902. Certainly papers on organophosphate toxins had been published in the international scientific press for decades and so there was good reason to believe the Allies had nerve gases of their own. The belief was reinforced by the fact that all mention of organophosphate toxins had disappeared from the Allied scientific press at the start of the war and the Germans assumed that this was due to military censorship. They were right, but the toxin the Allies were trying to play down was the insecticide DDT, which had been developed in Switzerland just before the outbreak of the war and was strategically important for military operations in malarial tropical regions. In his discussions with Hitler, Ambros stated that he could not believe that the scientists at the British chemical warfare establishment at Porton Down had failed to develop a similar nerve gas. How wrong he was! Ironically, the British actually discovered compounds applicable as nerve gases while experimenting with DDT just before the war, but they failed to recognise that the resultant substances were unusually deadly. The principle British researcher on such toxins, Bernard Saunders, did discover a nerve gas known as 'diisopropylfluorophospate' (DFP), but it was much less deadly than tabun or sarin and its true potential was never researched, let alone developed, by the British during the Second World War.

In any case, despite Ambros' assertions, Winston Churchill had made it very clear to Hitler that if Britain were attacked with poison gas, the British would saturate German cities with gas in retaliation. Certainly by this point in the war, the Allied strategic bombing force was much stronger than the Germans, the Allies were gaining air superiority over Germany, and Hitler had every reason to believe that if he were to use his 'victory weapon' on Britain, the Allies would strike back ten times as hard. Had Hitler known of his enemies' ignorance, the Second World War might well have taken a different course.

Ironically, Churchill himself almost gave the game away that Britain possessed no nerve gases. Although Britain had signed the Geneva Protocol, he had little squeamishness over the deployment of poison gases. To Churchill they were just another weapon. During the First World War and then in Iraq immediately after the war he had been so enthusiastic about chemical warfare that his wife jokingly called him the 'mustard gas fiend'.[35] Indeed, during the desperate days of 1940 when Britain was facing a German invasion, Churchill had energetically built up an arsenal of gas weapons to greet German troops if they landed on British shores.[36] Certainly, Sir John Dill, in his 1940 report, advocated launching immediate gas attacks if German soldiers set foot on British soil. Although Dill's report had met with stiff opposition he had all the support he needed from Churchill. Certainly, in Churchill's opinion, if the invading Germans managed to establish bridgeheads anywhere on the British or Irish coast, they should be hit immediately with gas, Geneva Protocol notwithstanding.[37] Within weeks, Britain's stocks of phosgene, which according to Greg Goebel had been discreetly shipped to Britain even before the United States formally entered the war,[38] and mustard gas were rushed to depots on the East Coast where they were loaded into spray tanks on specially adapted Lysanders,

Blenheims and Fairey Battles. The aircraft were then flown to airfields that were close to the beaches where the Germans were expected to land. In reality, these 'gas squadrons' were posted all the way down the East Coast of Britain from Lossiemouth on the Moray Firth to Kent. There is no doubt that gas would have been used if the Germans had invaded Britain or Ireland. In May 1941, the War Cabinet's Chief of Staff Committee agreed that if approval for the use of chemical weapons was ever given, 'The use of gas in Ireland [including Eire] would be ordered and controlled by the general officer commanding British troops in Ireland.'[39] Whether the Irish government of Eammon de Valera was ever informed that the British were prepared to spray the Irish coastline with powerful chemicals has never been revealed. Ultimately, although it was considered that the normal weapons of the three fighting services would almost certainly have succeeded in repelling an invasion, which of course they did, the importance of denying the invaders a lodgement was so overriding that it seems that any constraints on using poison gas would have been disregarded. The Germans certainly realised this:

> We had to reckon with the British, in the defence of their homeland...using every weapon and all means available...that might hold out even the slightest hope of success. We had to allow for the possibility of our troops being attacked...with non-persistent agents...as well as with vesicants.[40]

The incentives for the British to use gas against the invading forces would have been strong. With the country in mortal danger, the government would have been expected to authorise every means available that might have contributed to success, and, against massed troops struggling to establish beachhead positions, gas would have been highly effective. However, there were possible constraints on its use – most notably, the possible adverse effect on neutral opinion, especially in the United States, and the possibility of enemy retaliation in kind. The first of these could probably have been disregarded since although Britain had ratified the Geneva Protocol, the extremity of the country's position would certainly have justified a contravention of the protocol. The second constraint was stronger, although not nearly as strong as it was to become later in the war. German retaliation would have had to take the form of an aero-chemical attack against British cities. Certainly, at this stage in the war, the constraints against bombing civilian targets were still high, and Germany could have expected a far more hostile neutral reaction from gas bombing civilian objectives than Britain would have received from a gas attack against purely military targets.

Meanwhile, the plants making both phosgene and mustard gas went into overdrive. In response to Churchill's constant demands, the output of gas increased markedly. Once manufactured, the gas was stored in underground mines in North Wales. The statistics available show a steady increase and indeed by the middle of 1944 Britain had stockpiled more than five million gas-carrying shells and bombs. In the event, of course, the phosgene and mustard gas were never needed. The RAF stopped the Luftwaffe in the Battle of Britain and in October 1940 Hitler called off Operation Sealion (the invasion of Britain) and turned his attention to Soviet Russia.

Certainly the use of gas seemed a strong possibility in Soviet defensive actions against the German invasion of the Soviet Union. No information appears available on whether the Soviet forces considered using gas, but it is known that the German Army believed the probability that they would do so to be extremely high.

> The Russians did not use gas...In the Autumn of 1941 and the summer of 1942, we thought it possible that the Russians might employ gas...This was known to us from the instruction manuals we had captured shortly after the outbreak of war.[41]

There is, however, some evidence that Germany considered using gas against the Soviet Union,[42] and this is reinforced by correspondence between Stalin and Churchill in March 1942.[43] This subsequently led to Churchill's declaration in May 1942 that if Germany used gas against the Soviet Union, the United Kingdom would use its 'real and growing air superiority to carry gas warfare on the largest possible scale far and wide against military objectives in Germany'.[44] During the initial stages of the German advance, however, it seems highly unlikely that the German Army would have gained any significant advantage from gas: chemical warfare was ill-suited to the Blitzkrieg tactics used then. As the German advances were halted and finally reversed, different considerations came to the fore. In response to threats to German supply lines, posed by the growing partisan movement operating in the German rear, it was suggested that gas could be used to counteract this danger. However, by this stage in the war, as Allied dominance in the air grew, the possibility of Allied retaliation in kind had to be taken increasingly seriously.

In May 1942, German military authorities were reported to have stated that by mistake their forces had used gas once during the Polish campaign and again in the Crimea.[45] No further details were given, but a newspaper correspondent in Warsaw had reported on 3 September 1939 that mustard gas bombs had been dropped on the suburbs of the city.[46] The Polish government in exile also referred to the German use of gas against Warsaw.[47] These reports were complimented at the time by statements from Berlin that Polish troops had emplaced mustard gas landmines around a bridge on the outskirts of Jaslo, in Galicia, which had subsequently gassed fourteen German soldiers.[48] The report was built up into a fairly large propaganda operation in Germany with leaflets posted around the world and radio broadcasts, both of which alleged that the United Kingdom had supplied the mustard gas mines to Poland.[49]

With regard to the Crimea incident, the Soviet News Agency (TASS) had reported the German use of chemical mortar bombs on 7 May 1942.[50] During May 1942 the German forces in the Crimea also reportedly used poison gas against civilians sheltering in underground tunnels and catacombs.[51] 'The Germans sealed off all the exits and systematically introduced vast quantities of poison gas . . . five mass graves, with a total of more than 3000 bodies have been discovered in quarry galleries.'[52]

Even after the threat of invasion faded away, the British continued heavy production of chemical weapons. Indeed, as Table 4.2 illustrates, by far the biggest expenditure by the British government on research and development in the Second World War was on poison gas.

Table 4.2 Factories for munitions production

	Number of factories	Capital cost (£ thousands)
Explosives	8	10,444
Explosives materials	11	7,691
Small arms ammunition	5	9,106
Cartridge cases	10	5,350
Shells and fuses	7	9,106
Filling	6	15,867
Small arms	3	1,395
Signals and transport	5	999
Penicillin	3	1,946
Equipment and stores	2	76
Chemical defence	12	19,200

Source: Kim Coleman, op. cit., IG Farbenindustrie, PhD thesis.

By the middle of 1944, when the Allied armies had invaded France and were moving across Europe and the war was clearly moving towards its end game, Churchill astonishingly began to press once more for the use of gas. This time Churchill saw gas not as a way to defend Britain, but as a way to shorten the war. Certainly, Churchill was very strongly in favour of pounding German cities with gas bombs, and became more so in the summer of 1944 when the first V1 flying bombs indiscriminately killed and injured British civilians. On Churchill's list were 1600 'tactical targets' to be attacked with phosgene or mustard gas, including 60 cities with populations in excess of 100,000 such as Cologne, Frankfurt, Essen, Hamburg, Bremen, Munich, Nuremberg and Berlin.[53] The effect of an airborne gas attack on the people of these cities would have been incalculable. British military planning staffs investigated Churchill's new proposals and recommended against them.[54] It should be noted that their objections were not on grounds of humanity, however, but simply because the relatively crude gases available to the British would have required so many bomber payloads to have been effective that the conventional bombs then in use could do more damage. Apart from the dubious morality of Churchill's plan, Ismay and his staff saw no military advantage in spraying gas over a landscape across which Allied troops were expected to fight. They also foresaw a propaganda disaster of enormous proportions. Ismay replied to Churchill, 'On balance, we do not believe, that for us to start chemical warfare would have a decisive effect on the result or duration of the war against Germany.'[55] Churchill reluctantly gave up the idea.

As for the possible use of chemical weapons during operations in North Africa, climatic conditions accentuated one of the principal limitations of offensive chemical warfare, namely its dependence on the weather. No doubt, weapons intended for desert conditions could have been developed, but there was no great pressure to do so. The British, however, did maintain chemical warfare depots in Egypt, but the threat of enemy retaliation in kind against the United Kingdom provided an overriding sanction against their use.

The United States had never ratified the Geneva Protocol, but nevertheless, President Roosevelt considered poison gas a barbarous weapon. Indeed, he had no intention, unlike his British counterpart, of authorising its use, much to the disappointment of the American Chemical Warfare Service. The American chemical weapons programme only thrived because of the fear of Japanese chemical warfare efforts; indeed, American newspapers often printed reports of Japanese use of chemical warfare against the Chinese. Despite his reservations, Roosevelt issued a stiff warning that if the Axis powers used poison gas on American troops, they could expect massive retaliation in kind.

As revealed earlier, the Japanese did use chemical weapons in China before the Second World War but the newspaper reports that appeared in America during the war are hard to take at face value. Certainly, Chiang Kai-Shek wanted to encourage the Americans to continue to provide military assistance to the Chinese Nationalists, and stories of atrocities were an encouragement. Chiang Kai-Shek was also hoarding US military supplies and making little attempt to resist the Japanese, instead he was using his military stockpiles to deal with his rivals, the Communists. Claiming the Japanese used gas to win battles was a convenient excuse to win United States sympathy.[56] In fact, the Japanese had given up development and production of chemical weapons in 1941. Indeed, the Japanese stockpiles of poison gas were feeble compared to the mountain of chemical agents that the Americans had stockpiled, which actually exceeded, by a comfortable margin, the total amount of gas used by all sides in the First World War.

With so much gas stockpiled, accidents were likely to happen. On 2 December 1943, the merchantman, *SS John Harvey*, was awaiting its turn to be unloaded at the harbour of Bari in southern Italy. Unknown to most, the *John Harvey* was carrying two thousand 45-kg (100 lb) bombs full of mustard gas. Indeed, even most of the crew of the merchantman did not know of their deadly cargo. A few days earlier the Allied High Command had announced they had achieved complete air superiority over southern Italy. They had not informed the *Luftwaffe* however, and on the evening of 2 December 100 JU88 bombers swept into the harbour at Bari. The German raid was a stunning victory. Amongst other successful targets, the gas bombs on the *John Harvey* ruptured and as the ship sank a layer of mustard gas and oil spread over the harbour, while mustard gas fumes swept ashore in a billowing cloud reminiscent of the gas clouds in the First World War. Many civilians died as a result of these gas emissions and the officers of the *John Harvey* had already been killed as they frantically tried to scuttle the vessel. Sailors from the doomed ship were taken ashore and by the next morning 630 were blinded and developing hideous chemical burns. Within two weeks 70 of them had died. The British vessel, *Blisteria*, had picked up survivors during the raid but as a result of being in the vicinity, almost the entire crew went blind, some temporarily and some permanently, and many developed chemical burns. At first the Allied High Command tried to conceal the disaster for fear the evidence that gas was being shipped into Italy would convince the Germans that the Allies were preparing to use gas, and consequently provoke the Germans into pre-emptively using gas themselves. However, there were far too many witnesses to keep such a secret and in February 1944 the US Chiefs of Staff issued a statement admitting to the 'accident' and emphasising the United States had no intention of using gas except in retaliation to Axis gas attacks. In another incident a German shell hit an Allied gas dump on the Anzio bridgehead in 1944 and the gas started to drift towards the German lines. The Allied commander had to use the 'hot-line' technique to his opposite to convince him it was an accident and there was no intention to use gas.[57]

The Japanese, like the Germans, had little need of gas during their initial advances, but when they began to suffer reverses during the Islands campaigns, and as the United States' advance gathered momentum, it was perhaps inevitable that the Japanese Army should reconsider the use of gas. It is recorded that the Japanese Army General Staff sought permission to initiate chemical warfare during the Marianas Campaign, believing this was likely to be a decisive stage in the war.[58] General Tojo, however, turned down the army's request.[59]

For the US part, it is suggested that during the preparations for the Iwo Jima landings plans were made to use gas to spearhead the attack. The plans supposedly called for the island to be gas-shelled from offshore naval vessels, and the landings were to take place once the resultant ground contamination had cleared. The plan was approved by the US Joint Chiefs of Staff and by Admiral Nimitz, the theatre commander, but was rejected by President Roosevelt.[60] However, the authenticity of this plan has been strongly questioned.[61] The enormous losses at Iwo Jima strengthened the case of those who had argued for the United States to instigate chemical warfare. Nevertheless, gas was not used at either Iwo Jima or Okinawa, but the United States did continue to keep its possibilities at the forefront as the war with Japan went into its final stages.

A long-suppressed report, written in June 1945 by the US Army's Chemical Warfare Service, shows that American military leaders made plans for a massive pre-emptive poison gas attack to accompany an invasion of Japan.[62] The 30-page document designated 'gas attack zones' on detailed maps of Tokyo and other major Japanese cities. Army planners selected 50 urban and industrial targets in Japan, with 25 cities, including Tokyo, Osaka, Yokohama, Kobe and Kyoto, listed as 'especially suitable' for gas attacks.[63] In planning the invasion of Japan, America's military leaders expected the Japanese to fight with an almost fanatic fervour to defend their home islands and so the overall plan, code-named 'Operation Downfall', called for a two-stage invasion. First, 'Operation Olympic', an assault on the southernmost Japanese island of Kyushu scheduled for 1 November 1945 and, secondly, 'Operation Coronet', the assault on Tokyo scheduled for March 1946. In the first attack it was planned that bombers would saturate much of Japan with phosgene gas in the expectation that it 'might easily kill five million people and injure many more'.[64] Planners, however, also called for the use of mustard gas and AC in the preliminary to the second attack. Although public policy in 1945 was that the United States would only use gas in retaliation to a Japanese first use, it is clear that in private America's military leaders seriously considered striking first with poison gas. By the summer of 1945, US forces were already killing thousands of Japanese in fire-bombings. Given this, the step to killing with lethal gas was not a lengthy one. Certainly, as is a matter of public record, on 14 June 1945 Fleet Admiral Ernest King received a secret report from Army Chief of Staff, General George C. Marshall, and at a subsequent meeting the gas attack plan was approved. Certainly three days later, on 21 June 1945, orders were given to step up production of several types of poison gas to provide stockpiles in the massive quantities urged in the study. Senior staff, however, concluded that chemical warfare would only complicate the invasion of Japan and would not be a decisive weapon. In addition, co-ordinating and preparing America's allies for chemical warfare was also perceived as a major problem.[65] The use of the atomic bomb in August 1945 effectively ended the discussion.

Had chemical warfare been initiated on any of the occasions discussed in this chapter, it could have been said that there were sound military reasons, in the short term, for doing so. The possible effect of chemical warfare on enemy morale was certainly appreciated within military circles. In Germany the view was not only that the demoralising effect of gas was likely to be far greater than that of any other means of combat, but also that the effect on the human psyche was of greater importance than the ability of gas to produce casualties. The foundations of this doctrine had been laid in lectures given by Fritz Haber in 1924:

> All modern means of combat, although they appear intended to kill the enemy, actually owe their success to the intensity with which they affect the psychic stability of the enemy ... which in a decisive moment, induce the enemy to lose the will to fight and feel deflated ... Life in a trench subject to direct hit or cave-in is a terrific strain on human nerves, but the experience of war has taught us that the strain becomes tolerable ... Exactly the reverse is true of the

means of chemical warfare…Any change in the impressions felt by nose and mouth affects the psychic equilibrium through the unknown character of its effect, and is a new strain on the power of moral resistance of the soldier at a time when his entire strength should be devoted undividedly to his mission in combat.[66]

Haber was speaking principally about the battlefield use of gas, but his remarks could also be applied to the use of gas against civilians. It could be argued that given the sensational accounts of the effects of gas that had been appearing in the European press in the inter-war period, there was a considerable probability that a gas attack on a city would produce an effect on civilian morale out of all proportion to the weight of weapons used.

Three occasions are recorded when the notion of gas as a weapon of terror appears to have come near to being put into practice. The first was the proposal, already discussed, to arm V1 rockets with sarin. The second occasion was during the Allied encirclement of Germany in the autumn of 1944. Although the German military continued their opposition to the use of chemical weapons, their attitude was not unanimously shared in political circles. Pressure built up to use the large stocks of chemical weapons that Germany had by then accumulated, notably the nerve gas weapons. As noted earlier, Hitler was dissuaded from using nerve agents partly by Albert Speer. However, it should also be noted that during the summer of 1944 the overall control of Germany's chemical warfare capabilities had been transferred from Field-Marshal Keitel to SS-Obergruppenführer Brandt, Hitler's former physician. It is not, therefore, improbable that with this shift away from the army to the Nazi Party itself there was a corresponding shift in the balance of the incentives and restraints influencing the possible instigation of chemical warfare. The third occasion occurred immediately after the destruction of Dresden by British and US aircraft in February 1945. As the first news of this was received in Berlin, the initial reaction was to seek revenge. Among the alternatives considered was the demand by Goebbels that nerve gases be used against the British. It is not clear, however, how he proposed putting this into effect.[67]

It seems that a major constraint on the initiation of chemical warfare during the early part of the war was both a lack of the necessary material capability among the belligerents, and a general disinclination to acquire it. The cause of this apparent apathy can be found in the attitude towards chemical warfare in the military establishments on the eve of war. Certainly it is clear that in several of the nations who would soon be fighting in the Second World War the military value of chemical weapons was still undecided. Quite apart from the question of whether they themselves should be prepared to initiate chemical warfare, there was also the question of what their potential enemies were planning to do. No military establishment doubted the need for an efficient anti-chemical defence, and the Ethiopians' experience at the hands of the Italian Air Force had re-emphasised its importance. The reservation of the right to retaliate in kind had been made by some, but not all, of the major European powers when ratifying the Geneva Protocol. A retaliatory stockpile could then be regarded as a sanction of the treaty and an insurance measure. However, most leaders were still uncertain about the value of chemical weapons and, indeed, the overriding considerations in chemical warfare contingency planning were the intentions of the potential enemies. Nevertheless, it could be argued that by 1939 the major European nations all displayed signs of chemical warfare programmes and weapons procurement. Italy had given clear indications of this in Ethiopia. In Germany, opponents of the government had provided a spate of rumours, all later proved correct, that IG Farben was carrying out extensive chemical warfare research and development work for the German Army.[68] Germany knew that the Soviet Union had chemical weapons-manufacturing capability even if it only consisted of those factories Germany had helped to build in the 1920s. Germany also had information that the French and the British were carrying out chemical weapons trials in North Africa,[69] as indeed they were. Arguably, these instances were nothing more than an indication that the nations concerned were exploring possible retaliatory chemical warfare stances, but, in an increasingly tense international situation, many people must have been ready to believe that they indicated first-use intentions.

It seems clear then that at the end of 1939 each of the major belligerents suspected its enemies were prepared to initiate chemical warfare, whereas, in fact, none of them were willing to do so. Under the stimulus of these suspicions all were building up retaliatory stockpiles, but it was not until 1941 or 1942 that these stockpiles had grown to a point at which they provided a first-use option.[70] The fact of the matter was that gas was not a strategic weapon and, therefore, there was no incentive to expand the manufacturing base. Rear-Admiral Anthony Buzzard, a member of the British joint planning committees in 1943, recalled in 1968 that the Allies decided not to initiate chemical warfare mainly on the grounds that such an addition to the Allied strategic bombing campaign over Germany would not have been decisive. Interestingly, he added that had the Germans used gas against the Allies on the Normandy beaches, the result would have been decisive.[71]

In addition to the military constraints on initiating chemical warfare, there were also important non-military ones. In many countries during the 1920s and 1930s the question of chemical warfare capability was an issue of considerable public controversy. As a consequence, decisions about the use of chemical weapons during war had to take into account not only military considerations, but an assortment of political and non-military ones; questions of public opinion, international law and political expediency. In conjunction with the doubtful military value of chemical weapons in the light of their unpopularity with the general public, the financial restraints on their development during the years of the Depression and the taint of dishonour attached to chemical warfare, the acceptance of gas as standard weapon of war was seriously impeded. All these constraints served to hinder the acquisition and, therefore, the use of initiatory chemical warfare capabilities at the outbreak of war, and thus delay serious consideration on the employment of chemical weapons on later occasions when substantial military advantage could, perhaps, have resulted.

The failure of any combatant to use chemical warfare weapons during the Second World War remains an enigma. All the major combatants had large stocks of chemical weapons and some of the agents available in quantity were vastly superior to those used in the First World War. Indeed, most countries believed that chemical weapons would be used, and most had incentives to do so at one time or another. Reluctance to use such weapons out of distaste for them or fear of retaliation in kind played a part, but this was only one factor among several. It seems likely that the deciding factor was that the circumstances were never quite right to push any of the combatants over the threshold. In hindsight, however, it seems to have been a very close thing.

All armies learned several lessons from this non-gas war. The phrase 'Had Britain and the United States been prepared for war in 1936, there would not have been a war' was taken as a self-evident truth.[72] Certainly it was recognised that chemical warfare establishments, notably at Porton Down in the United Kingdom and Edgewood Arsenal in the United States, needed to be permanent organisations that concentrated on training, research and development, and chemical warfare preparedness. This lesson, from a slightly different angle, is reflected in the words of K.C. Royall, the US Under Secretary for War: 'The better job you do, the less likely it is you will have to put to actual use the products of your work.'[73]

The Allied armies began demobilisation activities almost immediately after the victory in Japan in August 1945, and by early 1946 chemical warfare personnel (now renamed chemical *defence*) numbered approximately the same as in the pre-war period. One contemporary observer commented, 'Gas warfare is obsolete! Yes, like the cavalry and horse-drawn artillery, it is outmoded, archaic and of historical interest only. This is the atomic age!'[74]

5
The Soviet Threat, Korea and Vietnam, 1945–1975

> Gas...was one of the most significant developments of the last war [First World War], but...has not been used in this war. The principal reason seems to have been that the power militarily ascendant at various times either had scruples against using gas or believed that his military ends could best be achieved without resort to it...We cannot be certain that in a future war an attacking power will be governed by similar scruples or conditions... Indeed, the emphasis on 'Blitzkrieg'...would encourage him to employ every means to achieve his end with speed and decision.
> – Tizard Report (February 1945)[1]

At the end of the Second World War, in accordance with the Potsdam Agreement which stated: 'All arms, ammunition and implements of war shall be held at the disposal of the Allies or destroyed', a large proportion of the chemical weapons which had been stockpiled, by both the defeated Axis powers and the Allies, were loaded onto old merchant vessels and scuttled off the coasts of Norway and Scotland. British dumping grounds for their own and captured German weapons included a 100-fathom site 20 miles off the coast of western Ireland and a site in the Bay of Biscay, both of which were used to disperse around 175,000 tons of weapons during the period from 1945 to 1948. The remaining British stocks of chemical weapons, about 25,000 tons, all manufactured during the war years, which included 6000 tons of tabun of German origin, were dumped during the period 1955–1957 at a 1000-fathom site in the Inner Hebrides.[2] Other German weapons, apart from those appropriated by the various Allied countries, were dumped in the Baltic Sea immediately after the war. Full records of these operations do not appear to have been kept, but it appears that there were at least three sites at which not less than 20,000 tons of weapons were dumped. One was in the Skagerrak off the coast of Norway where 20 ships whose cargoes included chemical weapons were scuttled by the British.[3] Another site was in the outer Bay of Kiel where ships that had been loaded with tabun weapons shortly before the war ended were scuttled under Allied supervision. Between 1959 and 1960, however, after mounting concern about pollution, the corroding ships were retrieved by the West German *Bundeswehr*, embedded in concrete and scuttled in the Atlantic off the Azores.[4] A third site was 20 miles off the East Coast of the Danish Island of Bornholm. Here, the Soviet Navy is said to have sunk a large number of captured mustard gas weapons by enclosing them in wooden crates, throwing the crates into the sea and sinking them with machine-gun fire.[5] Since then, fishermen and bathers in the Bornholm area have frequently suffered mustard gas burns.[6] Dumping grounds for French chemical weapons included a 1000-fathom site in the Bay of Biscay where 1700 barrels were sunk in 1965,[7] and it was also reported that the French dumped 24,000 tons of chemical weapons in the Mediterranean.[8] Sea dumps of Japanese chemical weapons notably comprised a mid-Pacific site at Wake Atoll where in 1968 the inhabitants were exposed to powerful airborne doses of chloropicrin,[9] and a site off the east coast of Chosi where, as in the Baltic, residents have suffered mustard gas injuries.[10]

However, despite what can be seen as a much publicised attempt to renounce chemical weapons at the end of the Second World War – weapons which, of course, had not been used – the Allies were already starting to argue among themselves over who should retain the secrets of the German nerve gases. The British were in no doubt about what should be done with the stocks of German chemicals which fell to their advancing forces from both the nerve gas arsenals and the factories where they were produced; most would be destroyed, but some supplies, mainly mustard gas and nerve agents, would be 'retained for possible use in the Far East'.[11] It was argued, 'on grounds of security' such stocks should be prevented from falling into the hands of the 'Russians or the French'.[12] In the event, it proved easy to keep the French from the nerve gas but over Russian acquisition the British had no control.

Despite German efforts to destroy evidence of research and development of tabun and sarin, the Russians captured the factory at Dyhernfurth intact. However, the Russians captured even more than this: they also took the nearly completed factory at Falkenhagen where the Germans were preparing to produce 500 tons of sarin a month. However, there were even more serious implications from the capture by the Russians of these two factories because, in addition to the secrets of tabun and sarin, the Russians also discovered the secrets of an even more poisonous nerve gas which the Germans had refined but not manufactured in quantity – soman. Soman, like tabun and sarin, was a colourless liquid in its pure form, with a slightly fruity odour. It worked in a similar way to tabun causing visual effects, runny nose, salivation, nausea, tremors, diarrhoea and involuntary urination and defecation, progressing ultimately to convulsions and respiratory failure. Soman, however, was twice as toxic as tabun by inhalation. There appears to be no documentation in the British archives concerning the Russian discovery of soman, so we can only guess at the reaction of the Allied scientific intelligence on finding that the Germans had discovered an even more potent nerve agent. Additionally, during the post-war interrogation of Professor Richard Kuhn (one of the German war chemists) the British discovered that all the documents relating to the work on soman had been captured by the Red Army and taken to Moscow.[13] While the British and American specialists were still analysing the nerve gases and attempting to isolate the specific mechanisms within the nervous system which were affected by them, the Russians possessed entire factories which could be operable within months. At the time, sources of information about the Russian capability for gas warfare were limited but at the end of the war the Americans concluded that the Soviet Union possessed a wide range of different gases. There were, they thought, probably 14 in all including First World War gases such as phosgene and mustard gas, in addition to the nerve agents. As the Cold War commenced, the belief that the Russians possessed this large chemical armoury was sufficient to ensure the survival of the chemical warfare establishments of the United States and Britain. Unlike the nuclear armouries of the superpowers, details of which were available, the exact size of the chemical arsenals were secret from the moment the Cold War began. In such a prevailing atmosphere of secrecy it was inevitable that suspicion would grow.

The Russians have divulged virtually nothing about their preparations for chemical warfare. Indeed, the only official statement that the Soviet Union even possessed chemical weapons was made before the Second World War began: 'Ten years or more ago, the Soviet Union signed a convention abolishing the use of poison gas [Geneva Protocol]...To that we still adhere, but if our enemies use such methods against us, I can tell you that we are prepared...to use them also.'[14] After this statement in 1938 the Soviet Union maintained a complete silence on its capacity for chemical warfare. Nevertheless, the formation of the Russian Army's Chemical Troops in the 1920s, who were consolidated and reorganised during the following decades, suggests the Russians were indeed seriously interested in chemical warfare. The testimony of a former Red Army officer, who defected to the West, which detailed gas training in the 1950s, further fuelled fears of a Soviet chemical attack on the West. However, Marshal Zhukov's address to the twentieth Party Congress, in Moscow in 1956, suggested that, in fact, the Soviets were expecting chemical weapons to be used against them by the West: 'Future war, if they unleash it, will be characterised by the massive use of...various means of weapons of mass destruction, such as atomic...and chemical weapons.'[15] During the late 1950s and early 1960s the Soviet double agent, Oleg Penkovsky, passed an enormous volume of intelligence material to the British and Americans, much of it about plans for chemical warfare. Penkovsky believed the Soviet Union was prepared to wage chemical warfare against the West but exactly what he told the British and Americans about Soviet plans for such warfare is not known, even today. During the mid-1960s the US Central Intelligence Agency (CIA) sponsored a book entitled *The Penkovsky Papers,* and according to this account Penkovsky told the British MI6 and the American CIA that in the Soviet Union there was a Directorate of General Staff who worked out methods of chemical warfare.[16] Undoubtedly, Penkovsky's information represented only a very small part of the Soviet Union's plans for chemical warfare but it was, nevertheless, a valuable source not least because it originated directly from Russia.

Many Western authorities believe that the Soviet Union invested heavily in chemical weapons during the 1950s as a cheap alternative to the tactical nuclear weapons that the West had developed and the Russians could not match. Even in the 1960s there was no evidence to suggest that the tons of mustard gas produced during the Second World War had been destroyed and, of course, it was known the Russians had the means and the expertise to produce nerve gases, tabun and sarin most definitely, and probably, soman as well. By the late 1960s the Russian array of chemical weapons was thought to range from Lewisite and mustard gas to rockets armed with nerve gas warheads.[17] In response to this perceived threat the West developed a range of weapons which must, to Moscow, have looked equally awesome.

The British had performed a series of experiments, mostly focusing on sarin, throughout the late 1940s and early 1950s, but they never went into full production of nerve gases. From the 1950s onwards, however, allegations periodically arose that British servicemen were used as human guinea pigs in nerve gas tests and in 2004 the British government launched a medical investigation into the health of 20,000 'volunteers' involved in chemical weapons trials at Porton Down. It is envisaged the two-year study will examine the death rates of all volunteers at Porton Down since 1939. Not only had the British experience in the First World War led to a historical reason for disliking gas weapons, but also the Second World War had exhausted Britain's financial resources. In such a climate there was increasing uncertainty about the future of the British chemical warfare establishment at Porton Down. Nevertheless, it could be argued that Porton Down's continued existence was down to the fact that although chemical warfare had not been used during the Second World War, and atomic weapons appeared to have eclipsed all else, the development of the nerve agents in Germany ensured that chemical warfare retained at least some part in the United Kingdom's military doctrine. It is perhaps difficult to appreciate the impact of these agents. Few earlier agents had been quite so insidious. The well-trained British serviceman was familiar with the characteristic smells of mustard gas, chlorine, phosgene, BBC and KSK – indeed few of the older agents were odourless. As the later stages of the First World War demonstrated, a few inhalations of these, at low concentration, before the gas mask was donned would do little harm. However, with the highly potent, odourless and colourless nerve agents that were able to exert rapid effects through the skin, eyes and respiratory tract, no such latitude was possible. It was, of course, impossible to demonstrate the actual effects of nerve agents on man and consequently difficult to build into service training any real protective measures. Unprotected men could not be put through gas chambers or allowed to see if nerve gases were decontaminated swiftly from their skin, if no effects were exerted. Unlike most of the older agents, the margin between mild effects at low doses and death at higher doses was small and, consequently, nerve agents could not be used in troop training. The problems of defence facing the British Chemical Warfare Service in the post-war years were, then, of a new kind. Equally, development of a UK chemical warfare capability based on nerve agents brought other problems in terms of future production by industry, weaponising, trialling and stockpiling. Military usage and the role of nerve agents had to be considered in detail. The advent of the nerve gases had, indirectly, emphasised the inadequacies of Porton Down's ageing facilities. Indeed, the massive lack of modern facilities and the effects of the ravages of wartime shortages, along with lack of repairs and upkeep, threatened the very future of the British chemical warfare establishment at Porton Down. In the subsequent era of post-war austerity and continued shortages, followed by defence cut after defence cut, the wonder is that the chemical warfare establishment survived at all.

As research on nerve agents progressed, the army and the Home Office, equipped with Porton Down's assessment of offensive potential and the hazards that might arise from chemical weapon use, could now issue their formal requests for the types of nerve agent-based munitions they required and for the new protective measures needed. On the offensive side, sarin emerged as the particular nerve agent on which the United Kingdom's chemical weapons were to be based. A series of bombs for the RAF was requested and the army requirement centred on a 25-lb shell and munitions for the 4.2-in. mortar. In the event, development work in the decade after VJ-Day (Victory in Japan) led to nothing because in 1957 the British government made the decision to abandon chemical weapons.[18] However, as John Pilger points out, this was not entirely true. Certainly chemical weapons were still being manufactured in Britain and sold to some 26 countries including Israel.[19] Indeed, in June 2002 the Department of Trade and Industry admitted that the sales of toxic chemical precursors (TCPs) had been authorised by the British government since 1957,[20] even though it was not known what they might be used for. In fact, the British government, under the terms of both the Geneva Protocol and the CWC, should only have licensed TCPs when it was 100 per cent certain they would not be weaponised. Professor Julian Perry Robinson, an expert on the CWC, said that a TCP such as dimethyl methylphosphonate could easily have been turned into sarin nerve gas, as used in the 1995 terrorist attack in Tokyo.[21] Nevertheless, in the 1950s, undoubtedly and most importantly, this chemical warfare excursion into the realms of offensive thinking and preparations gave Britain an excellent understanding of how other nations could use similar chemical weapons against the United Kingdom and its forces.

On the defensive side, service requirements were promulgated for a real-time detector for nerve agents in the field, shipboard detectors for Royal Navy vessels, prophylactics and therapy for nerve agent poisoning and a new respirator. The procurement cycle for some of these items was relatively quick, but for others, in particular the respirator, the process was one of continuous improvement reflecting advances in science and technology. However, to try to provide a succinct analysis of the first 20 post-war years of chemical weapon development in Britain is difficult mainly because many topics cannot be reported due to the non-disclosure of 'sensitive' documents by the British government.

Access to American sources presents no such obstacles and it is clear that the perceived Soviet threat after 1945 led to another chemical warfare build-up in the United States. Indeed, by the 1960s the United States had amassed a huge arsenal of chemical weapons and had begun production of a new nerve gas. In 1952, in an echo of Schrader's discovery of tabun, Dr Ranajit Ghosh working for a subsidiary of Britain's ICI discovered a new and deadly nerve agent while performing research into pesticides. The chemical was too dangerous to use as a pesticide and so ICI passed it to the British government, as Schrader had done in Germany in 1936 albeit under different circumstances. However, at this time, with limited financial resources, the British had already committed themselves to a pilot production of tabun and sarin and did not need, and could not afford, the research costs of a new agent. Consequently, in 1953, the British passed it on to the Americans. American scientists examined the new compounds and confirmed that a new series of nerve agents had been discovered that were more persistent and more toxic than tabun or sarin. This new series was designated 'V' series agents in 1955 because they were venomous in nature. Top priority was given to the investigation of these compounds and finally, in 1959, the Americans developed the formula into a weapon, designated 'VX' – the only significant nerve agent created after the Second World War and, indeed, the deadliest nerve agent ever created. A fraction of a drop of VX, absorbed through the skin, could kill by severely disrupting the nervous system.

In mild cases victims experienced a runny nose, eye pain and difficulty breathing; moderate cases caused increased eye symptoms, sweating, increased tightness of the chest and breathing difficulties, nausea, drowsiness, diarrhoea, headache and confusion; serious cases caused involuntary defecation and urination, twitching, staggering, convulsions, cessation of breathing and loss of consciousness followed by coma and death. The Annual Report of the US Army Chemical Corps for the year 1957 concluded: 'The reign of mustard gas, which has been called the King of Battle Gases since it was first used in July 1917, will probably come to an end.'[22] Although a cocktail of drugs (Atropine) could serve as an antidote, VX acted so quickly that victims would have to be injected with the antidote almost immediately to have any chance of survival. In addition to their work on VX, the Americans undertook other work in the post-war period on chemical warfare delivery systems, including artillery shells, the M-23 gas landmine, the M-55 unguided gas rocket and the MK-116 'Weteye' air-dropped gas bomb. In 1959 the first non-clustered bomb, the MC-1 750-lb sarin bomb was standardised. This was a modified general purpose demolition bomb that held about 215 lb of sarin filling. Although delivery systems for VX nerve agent were initiated during the 1950s, no system was standardised. In addition, many of the sarin delivery systems took longer to develop than planned and some were never standardised. Defensive systems were not ignored either, with the development of new gas masks, protective clothing, decontamination systems and kits, and primitive detection systems.

Another interesting development in the American research was the investigation of gases based on hallucinogens. In 1943, Dr Albert Hoffman, a researcher with the Swiss pharmaceutical firm Sandox, was investigating medicines derived from ergot, a fungus that infects wheat, when he spontaneously experienced wild hallucinations. Dr Hoffman had accidentally discovered the hallucinogenic drug 'LSD'. In the immediate post-war period American scientists wondered if hallucinogens might make effective 'humane' weapons that would not kill soldiers, simply eliminate their will to fight or, as it might have been put in a later era when hallucinogens became 'recreational drugs', persuade them to 'make love not war'. Consequently, during the mid-1950s experiments were conducted on volunteers, and controversially also on unwitting patients in psychiatric institutions, with mind-altering drugs. The results of these tests were encouraging, but LSD itself was not appropriate for military use, as it was too expensive to synthesise in volume and, additionally, was not a very good aerosol. The army finally found a substance named 'BZ' that was cheap to produce and could be dispersed in clouds over the battlefield. BZ, nicknamed 'Agent Buzz' for obvious reasons, made its victims ill causing them to vomit, stagger around and suffer memory lapses and hallucinations. During one test, according to a story, a soldier under the influence of BZ offered a second soldier, who was just as intoxicated, an imaginary cigarette. The second soldier turned him down saying it was the last in the pack![23] BZ was produced in pilot quantities, but then the army had second thoughts. It was too toxic and an enemy soldier on hallucinogens was just as likely to do suicidally insane and dangerous things as to become happy and agreeable. With this in mind, the army did not want to use such an unpredictable agent. However, the concept of BZ poses some interesting questions; most obviously, why did the Americans not develop a non-lethal agent that simply put enemy soldiers to sleep? In fact, the idea of 'knockout' gases had been around for a long time, but to develop an effective gas was not as easy or simplistic as it sounds. A gas could be made of opiates or some class of tranquilliser, but there would be no way to administer such a gas in a controlled fashion. Exactly what work American Army scientists did on knockout gases is not clear. What is clear is that the US Army never obtained them in any quantity. However, as became apparent in the October 2002 theatre siege in Moscow, the Russians did develop such gases in the Cold War period. Certainly the growing Soviet threat had concerned the chemical corps and the US Army throughout the 1950s and early 1960s and following Marshall Zhukov's address in 1956, in 1959 Major General Stubbs, the new Chief Chemical Officer, assessed the growing Soviet chemical threat thus: 'Soviet chemical weapons are modern and effective and probably include all types of chemical munitions known to the West... Their ground forces are equipped with a variety of protective chemical equipment and they are prepared to participate in large scale gas warfare....' He concluded: 'I believe I have given you enough to make you aware they pose a threat to the free nations of the world.'[24] Consequently, after much consultation with various groups around the United States on the need for a greater sense of urgency in attaining chemical preparedness, contending that to both military and civilian populations the threat of chemical warfare was as great as the threat of nuclear warfare, the 1961 experimental Project 112 was launched. Details of the American Cold War chemical warfare testing, Project 112, were only released at a news conference on 9 October 2002, where it was emphasised that these tests were authorised when there were 'serious and legitimate concerns about the Soviet Union's chemical warfare programme',[25] and that the tests were not conducted to evaluate the effects of dangerous chemicals on people but were operational tests of fighting capability.

Prior to the Second World War, it was primarily the European powers who used poison gas. After the Second World War 'old Europe' was increasingly replaced by the United States which, it can be argued, took on the policing of Third World countries, and such policing meant that chemical weapons were used in both Korea and Vietnam. Indeed, the United States had, for many decades prior to the outbreak of the Korean War in 1950, been the main producer, purveyor and user of chemical weapons. However, this was generally hidden from public view until the 1960s when the United States' use of chemical weapons in Vietnam was exposed.[26]

In June 1950, with the onset of the Korean War, the American Chemical Corps participated in its first military action since 1918. The Corps quickly implemented an increased procurement programme to supply the army with a chemical capability and defensive equipment. The new chief of the corps concluded that the need for such an ability was the number one lesson learned from the Second World War: 'It required the experiences of World War II to demonstrate that the most important basic factor in a nation's military strength is its war production potential and ability to convert smoothly and quickly its industry, manpower and other economic resources.'[27] Within a short time, however, the army's policy on chemical warfare and the lessons learned from the past were hotly disputed, particularly as the military situation in Korea changed. The action in Korea ultimately brought up the subject, at the highest level, of whether to initiate chemical warfare to save lives. The chemical corps commanders favoured the use of chemical weapons as humane weapons of war, particularly to offset the Korean's superior numbers. One officer stated the position quite bluntly: 'The use of mustard, Lewisite and phosgene in the vast quantities which we are capable of making... offers the only sure way of holding Korea at the present time. We are not playing marbles. We are fighting for our lives. Let's use the best means we have to overwhelm the enemy scientifically and intelligently.'[28] However, such ideas were apparently countered by fear that the Soviet Union would provide the Chinese and North Koreans with retaliatory chemical warfare materials.[29] Although neither side chose to initiate chemical warfare, there were allegations by the North Koreans and the Chinese that American forces employed chemical and biological weapons on the battlefield. Certainly, the American Chemical Corps did use riot control agents, including CS gas, to quell riots by prisoners of war,[30] although this was not admitted at the time. Also, certainly, in North Korea napalm and phosphorous bombs were systematically dropped in an effort to incinerate every city north of the 38th parallel.[31] The North Korean capital was a particular objective for Washington and on 11 July 1952 the US Air Force dropped 1400 tons of phosphorous bombs and 23,000 gallons of napalm on Pyongyang, levelling more than 1500 buildings and killing hundreds.[32] General Curtis LeMay described the devastation by saying: 'We eventually bombed every town in North Korea... and some in South Korea too.'[33] An official Chinese news agency report on 5 March 1951 stated that in the early afternoon of 21 February, two US aircraft dropped bombs charged with 'poison gas of an asphyxiating type' on North Korean positions 20 miles south-east of Seoul.[34] Other allegations of American use of chemical weapons in Korea are contained in a document prepared by a commission of the International Association of Democratic Lawyers (IADL). The report was entitled *Report on US War Crimes in Korea*, and was prepared after the commission visited Korea in 1952. The report referred to four alleged uses of chemical weapons in the Korean War. The first, and largest, was said to have taken place on 6 May 1951 when three B-29 bombers dropped mustard gas bombs over Nampo City, causing 1379 gas casualties of which 480 died of asphyxiation. The other incidents were said to have occurred on 6 July 1951 at Poong-Po Ri village, on 1 August 1951 at the villages of Yeng Seng Ri and Won Chol Ri, and on 9 January 1952 at Hak Seng village.[35] Consequently, it appears then that the American Chemical Corps ended the Korean War in a much stronger position than it had faced at the end of the Second World War. Major General Egbert F. Bullene, the new Chief Chemical Officer, summed up the Korean War and chemical warfare in general thus: 'Today, thanks to Joe Stalin, we are back in business!'[36]

During the 1950s the concept of warfare, and chemical warfare in particular, continued to change rapidly. The phrase that one could 'push a button' to start a war became popular, but in fact the lessons learned from the Korean War, the concept of a limited war fought without nuclear weapons against Soviet satellite states, not the real enemy, determined much of American and British planning. Indeed, in both the United States and Britain, the fact that now two wars had come and gone without the deployment of chemical weapons made it necessary for successive chemical research establishments to work continuously to remind their respective governments and their country that this might not be the case again. They strongly argued that the capabilities of the chemical warfare establishments constituted an insurance against the possibility of chemical attack in the future.

A growing guerrilla war in Vietnam soon made the US Army again re-examine its training programme, chemical warfare readiness and its no-first-use policy. As part of this sudden interest, the role of chemical weapons again came under intense scrutiny and debate. In 1963 one journalist stated: 'The best way for the United States to achieve its military aims in Southeast Asia would be to rely on chemical warfare.'[37] He then described how soldiers could 'sanitize' a particularly large area with gases and sprays that killed everything from vegetation to humans.[38] Then, in 1966, a retired US Army General suggested that mustard gas be used as an 'invaluable' weapon for clearing Vietnamese tunnels.[39] During this period other observers and authors also recommended revising the no-first-use policy, but most official histories promulgate that public opinion opposing the use of toxic chemicals was apparently the deciding factor against their deployment. However, in Vietnam the US Army did utilise defoliants and 'non-lethal' riot control agents in large quantities; the possible use of nerve gas is still shrouded in secrecy.

Chemicals were used against forested and agricultural lands in Vietnam as part of the US military strategy and tactics. This was the first time that chemicals designed to damage or kill plants had been used in war. The destruction of the land may seem a trivial thing in comparison to the human slaughter every war entails as to be of little concern. But when intervention in the ecology of a region on a massive scale occurs, an irreversible chain of events is set in motion, which continues to affect the agriculture of the area and therefore the people long after the war is over. The purpose of utilising such herbicides at this time, according to the US government, included 'To reduce the hazards of ambush by Vietcong forces.'[40]

In fact, chemical anti-plant agents had attracted military attention in the United States at the time of the Second World War. Indeed, in the words of a résumé published in 1946, 'Only the rapid ending of the war prevented field trials in an active theater of synthetic agents that would... affect the growing crops and make them useless.'[41] In all probability this was a reference to the planned use of anti-plant agents against Japanese rice crops.[42] US anti-plant agents were first used in war during the final year of the Korean War and then only on a very minor scale.[43] However, it was not until United States' involvement in Vietnam that they came to be employed on a significant scale in combat. In December 1961, President Kennedy authorised the Department of Defense to begin operational trials of anti-plant agents along certain lines of communication in South Vietnam,[44] a test programme known as Project Ranch Hand. In mid-1964 an expansion of Ranch Hand was authorised and in January 1965 approval was given 'to pre-strike targets with fighter aircraft, and to provide fighter escort for the spray aircraft'.[45] In October 1966, Ranch Hand was expanded again, especially in the area around Saigon and the Special Aerial Spray Unit was renamed the 12th Air Commando Squadron.

The US 12th Air Commando Squadron, in the first nine months of 1966 alone, defoliated a Vietnamese area of 1000 square miles, equivalent to the size of Derbyshire or the entire state of Luxembourg.[46] The amount of herbicides used is suggested by a 1967 newspaper item that highlighted that contracts for $57,690,000 of chemicals for defoliation had been awarded by the Defence Supply Agency.[47] The quantity of chemicals being purchased was not announced, but the amount in dollars suggests a purchase of between six and seven million gallons. Additionally, since prices remained stable, by 1968 the purchase would have been between seven and nine million gallons. At the same time, the value of British exports of herbicides rose from $730,986 in 1964 to $2,739,949 in 1967.[48]

In September 1966 *The New York Times* published a report that cacodylic acid, an organic arsenic containing compound, was also being used in Vietnam.[49] Cacodylic acid was a defoliant that was also toxic to man. According to the Merck Index, cacodylic acid was a dimethylarsenic acid containing 54.29 per cent arsenic, and was extremely poisonous.[50] Seventy grams would kill the average 150-lb man if administered subcutaneously.

Smaller doses would result in nausea, diarrhoea, headache, weak pulse and possibly coma. All these symptoms would flow from the paralysis of capillaries and degeneration of the lining of the intestinal tract, all known to be induced by arsenic poisoning. Eye witness reports from individuals in the sprayed areas, in particular the province of Tay Ninh, indeed described such symptoms. M.F. Kahn, an investigator for the International War Crimes Tribunal set up by Bertram Russell to investigate American war crimes in Vietnam, recalled an autopsy on a 5-year-old boy brought into the hospital in Tay Ninh suffering from abdominal pain and vomiting which was soon followed by collapse and death. The post-mortem examination revealed 'disseminated necrosis of the intestinal mucosa',[51] in other words, arsenic poisoning.

Official statements at the time, and to an extent even now, refer to only the less-toxic defoliants. However, Assistant Secretary of Defence, Cyrus Vance, when asked in 1965 whether arsenic and cyanide compounds were being sprayed over South Vietnam, replied, 'We are making limited use of them in the southern part of Vietnam but not yet in the north.'[52] It is hard to escape the conclusion then that aerial spraying of cacodylic acid continued unabated to at least 1971 and caused unknown casualties to human and animal life below.

Between 1961 and 1967 the scale of chemical anti-plant operations in Vietnam grew roughly in proportion to the overall involvement of US troops there. After 1967, however, there was a marked recession as countless pressures began to constrain the programme. There were four factors that can be seen to have contributed to this. First, the available commercial sources of anti-plant chemicals were becoming exhausted by the increasing military demand on them. Secondly, many people within the United States were becoming increasingly alarmed that anti-plant operations might undermine their agriculturally independent work by alienating farmers and other crop growers. This aspect had been emphasised by a study prepared by the RAND Corporation in 1967. Thirdly, the scientific community was expressing mounting concern that the anti-plant programme might permanently distort important sectors of the Vietnamese ecology. Finally, the view was expanding in the outside world that the combat use of anti-plant chemicals was contrary to the international laws of war.

The anti-plant agents favoured in Vietnam were 2,4-dichlorophenoxy-acetic acid (2,4-D), 2,4,5-trichlorophenoxyacetic acid (2,4,5-T), dimethyl-arsenic acid (cacodylic acid) and 4-amnito-3,5,6-trichloropicolinic acid (picloram). The formulation in which they were used is set out in Table 5.1.

Table 5.1   US anti-plant agents used in Vietnam

Agent	Active components of agent
Purple	n-Butyl 2,4-dichlorophenoxyacetate n-Butyl 2,4,5-trichlorophenoxyacetate iso-Butyl 2,4,5-trichlorophenoxyacetate
Orange	n-Butyl 2,4-dichlorophenoxyacetate n-Butyl 2,4,5-trichlorophenoxyacetate
White	Triisopropanolammonium 2,4-dichlorophenoxyacetate Triisopropanolammonium 4-amino-3,5,6-trichloropicolinate
Blue	Sodium dimethylarsinate dimethylarsinic acid

Source: Adapted from Stockholm International Peace Research Institute (SIPRI), Vol. 1, p. 172.

Purple and Orange were general purpose anti-plant agents used for the destruction of broad-leaved crops, such as banana, and for the defoliation of forest and brush growth. White was used for longer-term forest defoliation. Blue was a desiccant occasionally employed for rapid defoliation, but more usually for the destruction of rice crops. In the absence of detailed information on the purities of the active components used in each agent, the figures in Table 5.1 are approximate ones only derived by assuming 100 per cent purity and either calculating from the published acid-equivalent figures for each agent or, in the case where the specific gravity of the agent is known (Orange), calculating from the percentage composition.

The code names, 'Purple', 'Orange', 'White' and 'Blue', were derived from the colour of the stripe painted around the 55 gallon containers in which they were received from the United States. Agents Purple and Blue began to be used in Vietnam in 1961, but Orange gradually replaced Purple because of its lower volatility. Agent White came into use in 1966, at a time when Orange was in short supply. However, by the end of 1967, 90 per cent of the total agent sprayed was Orange (Table 5.2).

There are two reasons for a brief discussion here of napalm. First, the war in Vietnam led to an association of napalm with chemical weapons.

Table 5.2   Official US figures for consumption of anti-plant agents in Vietnam

Agent	1968	1969
Orange	2338	3269.5
White	2241	943.5
Blue	510	345.7

Source: SIPRI, Vol. 11, p. 173.

Second, one of the major issues in the development of chemical weapons is the role of secret weapons research in universities. The development of napalm is a fascinating case study of applied weapons research by a university chemist on his university campus.

Napalm is gelled petrol; in American terminology, gelled gasoline. Originally the term 'napalm' denoted the thickener that produced a gel when added to petrol; later it was broadened by usage to denote the incendiary gel itself. The name is derived from the first symbols of naphthenate and palmitate, two fatty acids first thought to be the active ingredients of the thickener. However, the material used in the original synthesis was mislabelled and actually contained the soaps of all the fatty acids of coconut oil, including lauric acid which was found to be essential to the gel. Although literally a misnomer, the name napalm was retained as the generic one for weapons of this type.[53]

Incendiary weapons have a long history in warfare and although the introduction of explosives in the fourteenth century temporarily eclipsed the use of incendiaries, the advantages of fire over blast were well known to military strategists. However, it was not until the advent of aerial warfare and the development of efficient incendiary substances, notably napalm, that fire reclaimed its role in war. Incendiary agents were used in the First World War, and an attempt was made by both the Germans and the Allied forces to use petrol in flame-throwers.[54] This was hazardous and generally unsuccessful, but it led to a recognition of the potential danger of incendiary warfare. The Treaties of St Germaine and Trianon in 1920 prohibited the use and manufacture of flame-throwers, along with that of chemical agents. However, it is of interest that the Geneva Protocol of 1925 that prohibited the use of poison gases did not deal with the use of incendiary weapons.

At the time of the outbreak of the Second World War the aid of Professor Louis Fieser of Harvard University, a distinguished organic chemist, was enlisted. The Japanese invasion of the East Indies had cut off supplies of rubber which was crucial to the thickener for napalm. The research programme that followed, conducted at Harvard University, saw the first successful napalm detonations on the games field behind the football stadium – an excellent example of applied weapons research in the universities.[55] The new napalm gel proved far superior to the original rubber-based gel, and napalm was used extensively by the United States in incendiary raids on Japan in the Second World War.[56] Napalm was also used in Korea where it was called the United States' 'best all round weapon'[57] and, of course, it was used extensively in Vietnam. SSg–E6 Tom M. Jackson (Vietnam, 12 September 1970 to 12 September 1971) described watching napalm set off from only 50 yards away:

It lit the night sky like...a volcano, all the hot and molten lava in the sky...like an oil well on fire...nothing but fire above you. It went over a hundred feet in the air above my location, burning the whole time. It lasted not even a minute but it seemed to last forever. Scared the hell out of me![58]

Napalm casualties were caused primarily by heat and carbon monoxide poisoning. The adhesiveness, prolonged burning time and high burning temperature of napalm favoured third-degree burns which were deep and extensive and often resulted in deformities. Another complication was often kidney failure and, worse still, the igniting agent in napalm weapons, white phosphorous, often became embedded in human tissue and continued smouldering and re-igniting long after the initial trauma. 'It sucks the air out of your lungs and burns them too, anyway it is not a good sight, a terrible way to die, no matter how it kills.'[59] It was reported in Korea that panic was more likely to be observed among napalm victims than among those wounded by other agents.[60] Additionally, it has also been claimed that napalm, used in bombing raids against Japan in the closing stages of the Second World War, caused more deaths than were caused by the atomic attacks on Hiroshima and Nagasaki.[61]

Although napalm and phosphorous are not in the strict sense chemical weapons, since they are mainly incendiary weapons, this classification is somewhat arbitrary. Napalm, when burning, gives rise to carbon monoxide – thus confirming reports of Japanese soldiers killed in the Second World War by napalm without visible burns. Many similar cases were observed in South Vietnam. Certainly, then, carbon monoxide intoxication greatly increases the lethality of napalm. Although certain incendiary weapons, such as flame-throwers, were coupled with chemical weapons when mentioned in peace treaties after the First World War and although napalm has, due to the circumstances in Vietnam, been coupled with chemical weapons, the most important document on chemical weapons in force in the 1970s, the Geneva Protocol of 1925, did not link the two. Policymakers at the time realised that any future measures for control and disarmament would be strengthened by clearly maintaining the separation of the two types of weapons, that is, chemical and biological on the one hand and incendiary on the other. Although explosives were also chemicals there had been no attempt to link them with chemical weapons in arms control documents. Although the distinction was a scientifically inexact one, policymakers realised that the special toxic properties of chemical weapons permitted their definition and control; it could be argued that policymakers did not make the mistake of blurring them by coupling them with incendiaries.

Soon after the March 1965 furore about the use of non-lethal gases in Vietnam, National Liberation Front (NLF) publications alleged that in addition to CS, CN and DM, the United States and south Vietnamese forces had employed a number of chemical casualty agents. On 5 April 1965, Hanoi Radio claimed that 10 weeks previously the United States had dropped 'lethal asphyxiating gases' similar to those used in the First World War on a hamlet in Phu Yen province. The gases were said to include adamsite, alpha-chlorocetophenone and tiphosgen.[62] It turned out that tiphosgen was a designation applied to CS, but by this time the use of agents such as CNS (CN and chloropicrin), VX and LSD was also being alleged.[63] Nevertheless, US officials consistently denied the employment of chemical casualty agents in Vietnam, including nerve gases and hallucinogens.

Despite the official denials concerning the use of gases in Vietnam, the use of CN, DM (adamsite) an arsenical compound, and CS (ortho-chlorobenzlmalonitrate) has now been admitted by the Americans. There is a persistent report, unconfirmed by the US military, that the hallucinogen BZ was used on at least one occasion, in Bong-San in March 1966. One other report, discredited and constantly refuted by the US government, is of the use of sarin nerve gas in Laos in 1970, the infamous Operation Tailwind. Following a damaging report concerning the use of sarin in Vietnam in *Time* on 15 June 1998, a Pentagon official told CNN and *Time* that the army, 'has found no documentary evidence to support claims that nerve gas of any type was used on Operation Tailwind'.[64] However, in 1970 the Chairman of the Joint Chiefs of Staff had confirmed the use of sarin in the Laotian operation and indeed in other missions to rescue downed US airmen during the Vietnam War.[65] More surprising, and potentially more embarrassing, Operation Tailwind had been launched to kill a large group of American GIs who it was believed had defected to the enemy. The actual preliminary raid to subdue the village lasted no more than 10 minutes and yet the body count according to the officer in charge was 'upwards of a hundred'.[66] With the camp destroyed, spotter planes ordered the Special Operations Group (SOG) Team to don the new advanced M17 gas masks.[67] Then came the explosions of the gas canisters. Mike Hagan, a veteran of the operation recalled: 'To me it was more of a very, very light, light fog. It was tasteless, odorless, you could barely see it.'[68] Unfortunately, some of the SOG team's masks had been damaged and some of the Americans began vomiting violently. Mike Hagan today suffers from creeping paralysis of his extremities, which has been diagnosed as nerve gas damage: 'Nerve Gas... the government don't want it called that... but it was nerve gas.'[69]

One month after the CNN/*Time* story a further article was published, *From the Wilderness*.[70] This article examined whether the CIA ordered the use of sarin not just in Operation Tailwind but in other covert operations during the Vietnam War too. The article suggested that there was a high probability that sarin was used not only against defectors but also against unwilling American prisoners of war whom the government decided would be a major embarrassment if they came home alive. Many of the distressing truths about American-ordered extermination of prisoners of war are documented,[71] but it was Scott Barnes' book about a failed 1981 prisoner of war rescue mission immediately followed by the alleged suicide of US Army chemical warfare and sarin gas expert, General Bobby Robinson that highlighted the issue.[72] Robinson was known to have been involved in moving sarin supplies in the area around Laos in 1970 supposedly planting sarin to blame the Soviets and thus motivate Congress to increase chemical warfare budgets. However, such operations were not unusual in covert operations and were hardly grounds for a suicide. Michael C. Ruppert cites a source who told him, 'It is much more likely that Robinson could have exposed the use of his sarin to kill Americans and he had to be killed – especially if he found out what his precious chemical agents were used for.'[73]

The US government easily discredited much of the CNN/*Time* story because the Pentagon said they had found no records of sarin use. Interestingly, however, following the *Time* disclosures and subsequent retractions, CNN set up an Internet bulletin board which was immediately bombarded by over 25,000 Vietnam veterans supporting the allegations concerning the use of sarin. The site was suddenly removed on 16 July 1998.

However, the American admission that they had used 'non-lethal' agents in large quantities in Vietnam caused a worldwide response that required the army to quickly explain the difference between lethal and non-lethal chemicals. First, the point was discussed in the context of the fact that the Geneva Convention did not prohibit 'riot control' and tear gases. Adamsite (DM), which was reported and admitted in use over the town of Hue in February 1968, was 'approved for use in... any operation where deaths are not acceptable' (for example, riot control) and, yet, in the field manual of the US Army it was stated DM may be used combined with CN in munitions and in 'military or paramilitary operations, in counter-surgency operations, or in a limited or general war... where deaths are acceptable'.[74] Interestingly, the use of DM by British forces at this time was ruled out on legal grounds and on the basis that its use would contravene the Geneva Protocol of 1925.[75] However, in the final analysis, there was no authoritative internationally accepted definition of the term 'riot control agents', for there was no international legal instrument incorporating any explicit reference to this category of weapons. Therefore, the only way to understand American policy at this time is to analyse the American definition, issued by the Pentagon on 1 February 1970, in which several connotations of the words 'riot agent' can be identified.[76] In the first place, they 'are *not* agents which result in prolonged incapacitation or death'. In the second place they have a 'temporary nature', their effects being 'not lasting' and 'dissipating quickly'. However, no comment whatsoever was presented to clarify the various terms used and so the ambiguity was perpetrated further.

The first information that such gases were being deployed in Vietnam came in March 1965, in *The New York Times*, when the Presidential Press Office stated, 'The gases are standard types of riot control agent.'[77] Later the same day the Secretary of State, Robert McNamara, stated the gases being used could easily be obtained through commercial channels and then added, 'Rather than use firepower, thereby jeopardising the lives of non-combatants, to drive the Vietcong out of the area . . . riot control agents were dispensed.'[78] As can be seen from these declarations, the use of these gases in shelters and caves was not under consideration, officially at least, in March 1965. Also, in March 1965 a news reporter in Vietnam noticed canisters of irritant agent in a helicopter in which he was travelling. After he enquired about these, the press agency for which he worked put out a story stating the US and south Vietnamese forces were 'experimenting with gas warfare'.[79] A US spokesman's views on the future of irritant agents were also quoted: 'Even if it does work, there is a real problem in getting it accepted . . . The ideas of it all brings back memories of World War I and mustard gas.'[80]

The following month *Time* stated, 'Compared with napalm bombs . . . or white phosphorus shell . . . the temporarily disabling gases in Vietnam seem more humane than horrible.'[81] There then followed many reports describing how the gases were now injected into tunnels and shelters including those where civilians were hiding.[82] Then came the story of Robert Bowtell in *The New York Times*: 'Non-toxic gas . . . being used against Vietcong Guerrillas, in tunnels north-west of Saigon, have killed one Australian soldier . . . Corporal Robert Bowtell, 21, of Sydney, died of asphyxiation although he was wearing a gas mask.'[83] This was the first official admission of the lethality of the so-called tear gases and it opened the floodgates.

Not surprisingly, stories such as these caused a frenzy in the outside world. Popular reaction was immediate and hostile. Newspapers gave prominence to accounts of the effects of exposure to DM and officials tried to counter these reactions by stressing the humanitarian possibilities of irritant agents: 'Under the circumstances in which gas was used in Vietnam, the desire was to use the minimum force required to deal with the situation to avoid death...to innocent people. We do not expect that gas will be used in ordinary military operations.'[84]

It appears then that the gases used by the United States in Vietnam were not simple harassing agents, owing to the way in which they were used in confined and closed areas where the concentration became high. The potential lethality of these gases was recognised, reluctantly and indirectly, by the United States in 1965. However, American statements on the use of chemical weapons in Vietnam have been consistent only in their evasiveness. Thus, even when they were trying to make the world believe they were using only 'tear gas' in Vietnam, it was revealed that DM was prohibited in US open air riot control because it was known that fatal casualties would result from its use.

On 24 March 1965, after the use of chemical agents in Vietnam had been discovered, the American Foreign Secretary told the press:

> We are not embarking upon gas warfare in Vietnam. We are not talking about agents or weapons that are associated with gas warfare in the military arsenals of many countries. We are not talking about gas that is prohibited by the Geneva convention of 1925 or any other understandings about the use of gas.[85]

After this statement different commentators frequently repeated the same line. Finally Congressmen Kastenmeier pointed to the fact that this interpretation was 'invented' by the American government after the use of chemical agents had been presented as a *fait accompli* of chemical warfare, and the Pentagon statement had been issued as a reply to public reaction.[86] The matter became a point of dispute among the states party to the Geneva Protocol, prompted by Australia, itself both involved in the Vietnam War and the use of riot control agents there, when it declared: 'It is the view of the Australian Government that the use of non-lethal substances such as riot control agents, herbicides and defoliants does not contravene the Geneva Protocol.'[87] Before the matter of interpretation of the protocol became acute in 1969, the United Nations had reacted to the first news about the use of riot control agents in Vietnam by means of a Resolution, which called for strict compliance by all states with the principles and objectives of the protocol (Table 5.3). Australia demanded that the term 'chemical weapons' be specified. When this was not done, and after the phrase 'principles and norms' had

Table 5.3   Riot control weapons in use in Vietnam in 1969

Code number/kind	Delivery system	Payload (gms)
ABC-M-7A3CS hand grenade	Hand or rifle	11.5
ABC-M-25AICN-1 hand grenade	Hand	50
ABC-M-25A-CN-1 or CS-1 hand grenade	Hand	50
M-3CS-1 or CS-2 agent disperser	Portable	4000
M-4CS-1 or CS-2 agent disperser	Helicopter or vehicle	22,000 per hopper
M-5CS-1 or CS-2 agent disperser	Helicopter	18,000 per hopper

Source: Adapted from SIPRI [406], Vol. 1, pp. 192–193.

been replaced by the phrase 'principles and objectives',[88] it was possible even for those at whom the Resolution was in fact levelled, the United States and Australia, to vote in favour of it. This was the closest the United States had ever come to ratifying the Geneva Protocol of 1925.

> And they assembled them at a place called Armageddon. The seventh angel poured his bowl into the air, and a great voice came out of the temple, from the throne saying, 'It is done!' and there were great flashes of lightning, loud noises, peals of thunder and a great earthquake such as had never been seen since men were on the earth, so great was the earthquake.[89]

According to Christian belief these verses from the Bible are a direct reference to nuclear war – the kind of war expected in the Cold War period. The sixteenth-century prophet, Nostradamus, made reference to another kind of attack in one of his extracts: 'The sky will burn at 45 degrees. Fire approaches the city. In an instant huge scattered flame leaps up.'[90] Was this then a reference to napalm and chemical warfare – the reality of the Cold War period?

6
The Middle East, Afghanistan, Bosnia and the Gulf

While the United States was still involved in the Vietnam War, another war in the Middle East brought the subject of chemical warfare back to the forefront. From 1963 reports began to filter out of the use of poison gas in the Yemeni Civil War. It was alleged that the gas was being used by the Egyptians on behalf of their allies, the Yemeni Republicans. The charges came from the Yemeni Royalists, their supporters, the Saudi Arabians, from journalists and finally, most authoritatively, from the International Red Cross.

The first suggestion of the use of gas in the Yemen came on 8 July 1963, with an article in the *Daily Telegraph* stating there had been a gas attack on the village of Al Kannna that killed seven civilians.[1] Immediately the United Nations responded by investigating the allegation, sending an observation team to Yemen, but their report concluded there was no evidence of a chemical attack.[2] Further newspaper articles described attacks taking place between 1963 and 1967 although most disagreed on dates, locations and effects of the attacks. Consequently, the world took little notice. However, in 1966 Wilfred Thesiger, a British archaeologist, came to a village shortly after it was alleged to have been subject to a gas attack. He reported seeing 20 victims of what he termed 'blinding gas', and said he treated a boy suffering from 'blister gas' that might have been a form of mustard gas.[3]

Much like the progression of chemicals in the First World War, the Egyptians allegedly started with tear gases meant to terrorise rather than kill, then progressed to mustard agents which caused more serious casualties, and finally to nerve agents which were meant to kill large numbers quickly. The combination of the use of nerve agents in January 1967, and the outbreak of war between Egypt and Israel in June 1967 finally attracted world attention to the events in Yemen.

On 5 January 1967 an attack occurred on the Yemeni village of Kitaf, the military headquarters of the Yemeni Royalists. During an air raid, bombs were dropped upwind of the village and produced a grey-green cloud that then drifted over the village.[4] According to newspaper accounts,[5] 95 per cent of the population up to 2 km downwind of the impact site died within 10–50 minutes of the attack. Another attack was reported to have taken place on the town of Gahar in May 1967 and additional attacks occurred, also in May, on the villages of Gabas, Hofal, Gadr and Gadafa, killing over 243 citizens.[6] Shortly after these attacks the International Red Cross examined victims, soil samples and bomb fragments and officially declared that chemical weapons, identified as mustard gas and possibly nerve agents, had been used in Yemen.[7] The salient points in this statement were that poison gas had been used in violation of the Geneva Protocol, that it was used against civilians and that nerve gas might have been employed in some of the attacks. However, reports of possible chemical use in certain parts of the world, particularly those inaccessible to official and technical observers, were difficult to confirm, and therefore condemn, without accurate and verifiable information. During the Yemen Civil War news reports alone proved informative but unreliable, and samples taken from the scenes of the alleged attacks apparently did not lead to further political or military action. Nevertheless, those responsible for the allegations of chemical warfare have advanced a number of explanations as to why chemical weapons were employed during the war. First, gas was viewed as a means of neutralising enemy strong-posts in mountain caves that were invulnerable to conventional attack.[8] Secondly, gas was seen as an efficient means for coercing tribesmen whose allegiances were vacillating between the warring parties. Radio Sana'a frequently broadcast warnings that any village that went over to, or gave support to, the Royalists would be gas-bombed.[9] Finally, the Yemeni Civil War provided an attractive proving ground for experimental chemical weapons.[10] The validity of these explanations is, of course, no greater than the authenticity of the reports to which they refer. Perhaps, most importantly, with the world distracted by the Arab–Israeli Six-Day War and events in Vietnam, politics discouraged a universal condemnation and follow-up response. In effect, the world powers let the event pass much as they had when Italy used chemical warfare agents against Ethiopia in the 1930s.

The 1967 Arab–Israeli Six-Day War has been described as the war that came very close to being the first major war where both combatants openly used nerve agents and chemical warfare. Fearing a pending attack from its Arab neighbours, on 5 June 1967 the Israelis launched a pre-emptive strike against Jordan, Egypt and Syria. This action included an invasion of the Sinai Peninsula, Jerusalem's Old City, Jordan's West Bank, the Gaza Strip and the Golan Heights. Reports soon appeared that the Egyptians had stored artillery rounds filled with nerve agents in the Sinai Peninsula for use during a war. The Israelis, perhaps reflecting on Egypt's possible testing of the weapons in Yemen earlier in the year, realised that their troops and cities were vulnerable to attack. The fact that chemical weapons were not used during the Six-Day War was possibly due to Israel's pre-emptive action or possibly due to the newspaper reports of the Yemen Civil War. Certainly, the Israelis felt threatened enough to place frantic orders for gas masks with Western countries. However, this last-minute call for gas masks and nerve agent antidote came too late to have prevented enormous casualties had the Egyptians deployed nerve gas. On the other hand, the Egyptians claimed Israel was preparing for biological warfare.[11] A United Nations-sponsored ceasefire ended the fighting on 10 June 1967, and the potential chemical war did not occur.

While concern over the potential and actual use of chemical agents grew during the 1960s and 1970s, the United States continued its chemical warfare production programme. The first, and only, incapacitating agent (excluding riot control agents) standardised was completed in 1962 – 3-quinuclidnyl benzilate (BZ) was a solid but was disseminated as an aerosol. The major problem with this chemical agent for military purposes was its prolonged time of onset of symptoms. The estimate was 2–3 hours before the enemy would become confused and therefore vulnerable. This was a tremendous disappointment to military strategists who were hoping for a quick-use, non-lethal agent. A second problem was its visible cloud of smoke during dissemination, which limited the element of surprise.[12]

Having concentrated on nerve agent bombs during the 1950s, the US chemical warfare establishment turned its attention to artillery, rocket and other delivery systems during the 1960s. Indeed, in 1960 the first nerve agent landmine was developed. This mine resembled the conventional high-explosive landmine, but it held 11.5 lb of the chemical agent, VX. The early 1960s was the peak of the nerve agent rocket programme, a programme first started at the end of the Second World War to duplicate the German V2 missiles used against the United Kingdom. For short-range tactical support, the M55 115-mm rocket was developed. This rocket was loaded with 11 lb of VX or sarin and its range was over 6 miles. Each launcher held 45 rockets that could be fired simultaneously. For middle-range tactical support, the 'Honest John' rocket was developed. This rocket had a range of 16 miles and the warhead held three hundred and fifty six 4.5-in. spherical bomblets, each containing 1 lb of sarin. The first long-range warhead had a range of 75 miles and the warhead contained 330 sarin bomblets. During the 1960s more developmental projects added chemical warheads to other long-range missiles, such as the Pershing missile, which had a range of over 300 miles. By the end of the 1960s the United States had a truly awesome chemical warfare capability.

However, in the late 1960s growing protests over the US Army's role in Vietnam, the use of defoliants and the resort to riot control agents both in Southeast Asia and to quell student demonstrations on the home front gradually increased public hostility towards chemical weapons. Such controversies helped to keep the fact that the United States had large stockpiles of lethal chemical weapons an embarrassment and so with world opinion strongly against such weapons, there was consequently no way the Americans could use poison gas except in retaliation. Generally, opinion was that because the United States had the nuclear deterrent, the need for lethal chemical weapons was arguable. Under growing pressure worldwide, in April 1969, the Secretary of Defence tried to explain the US chemical warfare policy. In part, he stated:

> It is the policy of the United States to develop and maintain a defensive chemical and biological capability so that United States military forces can operate for some period of time in a toxic environment if necessary; to develop and maintain a limited offensive capability in order to deter all use of chemical . . . weapons by the threat of retaliation in kind; and to continue a programme of research and development in this area to minimise the possibility of technological surprise.[13]

The explanation did not help. In July, the United Nations issued a report on chemical and biological weapons that condemned production and stockpiling of WMD. Congress stepped in and on 11 July 1969, for no apparent reason, it revealed the army was conducting open air testing of nerve agents at Edgewood Arsenal and Fort McClellan. Shortly after this disclosure, more than 100 protesters were at the gates of Edgewood Arsenal and three days later, giving way to the pressure, the army announced the suspension of the tests. Immediately a committee was put together to conduct a safety review, but the positive publicity of this new committee was soon forgotten when the army revealed that they had also conducted nerve agent testing in Hawaii between 1966 and 1967, something they had previously denied.[14] Under increasing public pressure, in November 1969 President Nixon took action against chemical warfare. First, he reaffirmed the no-first-use policy for chemical weapons, and secondly, he resubmitted the 1925 Geneva Protocol to the US Senate for ratification. President Nixon concluded by elucidating his future hopes: 'Mankind already carries in its own hands too many of the seeds of its own destruction. By the examples that we set today, we hope to contribute to an atmosphere of peace and understanding between all nations.'[15] These actions were effective in stopping the production of chemical weapons in the United States until 1979.

Although President Nixon had called in 1969 for the ratification of the Geneva Protocol, it was not until 1974 that it was finally ratified through the US Senate and President Ford officially signed it on 22 January 1975. He did, however, exempt riot control agents and herbicides from inclusion in the agreement.

In 1973 another war quickly brought chemical warfare preparedness back to the forefront. The Arab–Israeli Yom Kipper War lasted from only 6 to 24 October 1973, but the ramifications for international chemical programmes lasted much longer. It is generally thought that Israel initiated a chemical warfare programme in the mid-1950s, which focused on studies of chemical-incapacitating agents which were designed not to kill but 'incapacitate' an adversary for a certain amount of time. What is not certain, however, is whether the Israelis were involved in an extensive effort to identify practical methods of synthesis for nerve gases, tabun, sarin and VX.[16] Karel Knip's findings that such research was undertaken from about the mid-1950s is constant with other indications that Israel did indeed develop a coherent chemical warfare programme.[17]

Following the war, the Israelis analysed the Soviet-made equipment they had captured from the Egyptians and Syrians. They discovered portable chemical proof shelters, decontamination equipment for planes and tanks, and that most Soviet vehicles had air filtration systems on them to remove toxic chemicals. Another item of note was a chemical agent detector kit that was designed to detect nerve, blister and blood agents. US specialists later determined it could detect even low concentrations of nerve agents, mustard agent, phosgene, cyanide and Lewisite. Overall, the experts reported finding sophisticated chemical defence material and a 'superior quantitative capability for waging chemical war'.[18] The indications were that the Soviets were ready for extensive chemical warfare and might actually be planning to initiate chemical warfare in a future war. While Soviet secrecy kept the details hidden, the Soviet Union undoubtedly engaged in a chemical arms build-up that almost certainly matched that of the Americans in the 1960s. The Soviet Union seemed to prefer soman (GD) to sarin, and were believed (later verified) to have stolen the formula for VX, and additionally developed a VX variant that remained effective in extreme cold. It was clear then, in the early 1970s, that the Red Army possessed a strong chemical warfare capability. Certainly, there is no doubt that the Soviet Union continued to produce VX, sarin, soman and Lewisite between 1945 and 1987.[19]

The combination of the findings of sophisticated chemical defence material, and therefore the Soviet Union's capacity for waging chemical war, along with the decline in the United States' chemical warfare programme led in 1976 to the Secretary of the army reversing the 1969 decision to abolish the American chemical warfare programme. Nevertheless, in 1977 the United States began new efforts to reach an agreement with the Soviets on a verifiable ban on chemical weapons. This effort was unsuccessful and as a result the American chemical programme recommenced in 1979.[20]

The end of the chemical weapons production programme in 1969 had stopped production but left one type of chemical retaliatory weapon still in development. Back in 1950 the US Army had begun looking at binary weapons. Until that time chemical weapons were unitary chemical munitions, in other words, the chemical agent was produced at a plant, filled into the munitions and then stored ready to be used. Since most of the chemical agents were extremely corrosive, unitary munitions were logistical nightmares for long-term storage. The binary concept was to mix two less-toxic materials and thereby create the chemical agent within the weapon after it was fired or dropped. Because the two toxic materials could be stored separately, the problem of long-term storage and safe handling of chemical weapons was solved. However, it was only after the production of unitary chemical munitions was halted in 1969 that the binary programme began to receive more priority. Eventually, the United States built a 155-mm artillery shell to deliver binary nerve gas in the form of GB and sarin. The shell contained two chambers, one filled with methylphosphoric difluoride (DF) and the other containing simple isopropyl alcohol. When the shell was fired, the barrier between the two chambers broke, and the rapid spin of the shell mixed the two precursors to form the gas, which was then dispersed when the munition burst. VX could also be produced in this way by mixing VC and sulphur. Despite these developments, and despite the ever-growing suspicion of Soviet intentions in the mid-1970s, the US Congress refused to fund the programme.[21]

Starting in about 1975, reports of the use of chemical agents in various small wars in Southeast Asia and Afghanistan began to attract international attention. Interviews with Hmong villagers in Laos suggested that Vietnam and Soviet forces might have used chemical and toxin weapons against these people. Villagers spoke of aircraft pouring out 'yellow rain' that caused choking, chemical burns, massive bleeding and rapid death.[22] There were many reports, but the combination of symptoms recounted resembled the action of no known chemical agent and it was therefore suspected that the 'yellow rain' had represented some mix of chemical agents and a new chemical or biological toxin. The evidence was thin and although some deadly mycotoxins (fungal poison) were discovered, they were nowhere near as toxic as any nerve gas and much more expensive to produce. In the absence of any definite information 'yellow rain' was dismissed as an unsettling rumour.

However, the suspicions continued to grow. In 1978, reports from Kampuchea claimed the Vietnamese and their allies had killed over 980 villagers using chemical weapons. Even prior to the Soviet invasion of Afghanistan in December 1979, reports were already circulating that Soviet troops were using chemical weapons against the mujahidin soldiers. However, although the mujahidin spoke of 'nerve gas', they described clouds of coloured smoke and choking symptoms which would suggest they were subjected to asphyxiants rather than nerve gas which would have generally been colourless and caused convulsions. The reports were never confirmed. It does seem plausible that the Soviet Union used riot agents in Afghanistan, and riot agents can be lethal in high concentrations. Nevertheless, the reports from Afghanistan, as well as the 'yellow rain' stories from Laos, still provided little real evidence of any serious use of chemical weapons. However, in 1982 the Soviet Union attempted to legitimise their use of chemical weapons by saying that although they had signed the Geneva Protocol (in 1928), Laos, Kampuchea and Afghanistan were not signatories and therefore the Soviet Union's actions against them were justified. Intelligence sources in the United States believed the Soviet Union considered most toxins to be chemical agents and, if this were the case then, the Soviet Union would be permitted under the Geneva Protocol to use them in retaliation or against non-signatories. All in all, the possible use of chemical weapons by the Soviet Union was, nevertheless, taken as an indication that they were still continuing an active chemical warfare programme.

Throughout the 1980s the United States monitored the war in Afghanistan, often thinking of it as the Soviet Union's Vietnam. Despite denials by the Soviet Union the United States finally went public with charges that chemical warfare had been used in Southeast Asia and Afghanistan in 1980. However, problems with the collection of samples and the remoteness of the sites prevented definitive evidence from being obtained. In 1982 the US Secretary for State, Alexander M. Haig Jr, presented a report, entitled *Chemical Warfare in South East Asia and Afghanistan*, to the US Congress. In this report, after describing the evidence, he concluded:

> This evidence has led the United States Government to conclude that Laos and Vietnamese forces operating under Soviet supervision, have...since 1975, employed lethal chemical and toxin weapons in Laos; that...Vietnamese forces have...since 1978, used lethal chemical agents in Kampuchea; and that Soviet forces have used a variety of lethal chemical warfare agents including nerves gases, in Afghanistan since...1979.[23]

Based on the evidence provided by Alexander Haig, senior defence department personnel concluded that the Soviet Union possessed decisive military advantage because of its chemical capabilities. The Haig Report, however, was not so enthusiastically received by the rest of the world, based as it was at best on flimsy evidence, and indeed it failed to galvanise world opinion because much like the situation in the Yemen Civil War the United States was unable to prove beyond a shadow of a doubt that chemical agents and toxins had been used in Southeast Asia and Afghanistan. Nevertheless, one American military writer summed up the general American view thus: 'The probable use of chemical weapons by Soviet forces in Afghanistan is...significant...Afghanistan is proof positive that the Soviets do not consider these devices as special weapons. Considerations of utility and not morality will govern Soviet use of them in a future conflict.'[24] Despite the possible use of chemical weapons, the Soviet Union was unable to win the war and in December 1988 was forced to meet with rebel forces to discuss a withdrawal of Soviet troops from Afghanistan. In June 1989 the final withdrawal was announced and completed a month later. By that time, however, the Soviet Union was not the only country feared in the West in terms of chemical weapons capability, for there was increasing widespread suspicion that lesser states with militant and authoritarian regimes were developing chemical and biological weapons as a military equaliser.

The beginning of a war in the Middle East eroded the high status previously given in the United States to the Soviet threat. On 22 September 1980, the armed forces of Iraq launched an invasion against their neighbour, Iran. The invasion struck Iran when the Islamic revolution that overthrew the Shah's regime in 1979 was still consolidating its hold. The Iraqi Army, trained and influenced by Soviet advisors, had organic chemical warfare units and possessed a wide array of delivery systems. When neither side achieved dominance, the war quickly became a stalemate. As in the First World War, in an effort to break the stalemate, the Iraqis employed their chemical agents as an offensive measure against the much less well-prepared Iranian infantry. The first reported use of chemical weapons occurred in November 1980 and throughout the next few years additional reports and rumours circulated of new chemical attacks.

In November 1980, Tehran Radio broadcast allegations of Iraqi chemical bombings at Susangerd. Three and half years later, by which time the outside world was listening more seriously to such charges, the Iranian Foreign Minister told the Conference on Disarmament in Geneva that there had been at least 49 instances of Iraqi chemical warfare attacks in 40 border regions. He announced that the documented dead totalled 109 people, with hundreds more wounded.[25] This statement was made on 16 February 1984, the day on which Iran launched a major offensive on the central front. According to official Iranian statements, during the time following the Foreign Minister's allegation, Iraq used chemical weapons on at least 14 further occasions, adding more than 2200 to the total number of people wounded by poison gas.[26]

One of the chemical warfare attacks reported by Iran, at Hoor-ul-Huzwaizeh on 13 March 1984, has since been conclusively verified by a team of specialists sent to Iran by the United Nations Secretary-General. The evidence in their report lends substantial credence to Iranian allegations of Iraqi chemical warfare on at least six other occasions during the period between 26 February and 17 March 1984. On 7 March the International Red Cross reported on 160 cases of wounded soldiers in Tehran hospitals who presented a clinical picture which led to the presumption of the recent use of substances prohibited by international law.[27] Two days later the US State Department announced that the US government had concluded that all the available evidence indicated that Iraq had used lethal chemicals.[28] Iraq denounced the American statement as 'political hypocrisy', 'full of lies' but at the same time the general commanding the Iraqi Third Corps spoke as follows to foreign journalists: 'If I have to finish of the enemy and if I am allowed to use chemical weapons, I will not hesitate to do so.'[29]

On 30 March 1984, the United Nations Security Council issued a statement condemning the use of chemical weapons during the Iran–Iraq War. The same day the US government announced it was instituting a special licensing requirement for exports to Iran and Iraq of particular chemicals that could be used in the manufacture of chemical weapons. Other governments took similar steps. Following this statement, reports of Iraq's use of chemical warfare agents dwindled, but they did not stop altogether. For example, a British television team filming on the Iranian side encountered evidence of a mustard gas attack in mid-April 1984, but the Iranian media did not publicise this report, perhaps mindful of a potential negative impact on their domestic audience.

From an unexploded bomb found at an Iraqi attack site, the United Nations were able to establish the use of mustard gas. Later published findings by the United Nations ascertained that the mustard gas had been manufactured using thiodyglycol as the starting material. This was not the favoured method of production for either the United States or Great Britain, and so the finger of suspicion was pointed towards the Soviet Union. The second poison gas, also identified by the United Nations team was the nerve gas tabun.[30] In conclusion, the United Nations investigators pointed out the dangers of this chemical warfare:

> It is vital to realise that the continued use of chemical weapons in the present conflict increases the risk of their use in future conflicts . . . In our view, only concerted efforts at the political level can be effective in ensuring that all the signatories of the Geneva Protocol of 1925 abide by their obligations. Otherwise if the Protocol is irreparably weakened after 60 years of general international respect, this may lead, in the future, to the world facing the spectre of the threat of chemical weapons.[31]

In a sense, then, the Iran–Iraq War reintroduced chemical warfare on a large scale shattering any belief that the use of such weapons had been successfully contained by international agreements. By deploying mustard gas and tabun the Iraqis broke a taboo and therefore made it easier for future combatants to find justification for chemical warfare. Nevertheless, despite Iraq's use of chemical weapons the war failed to reach a military conclusion and in August 1988, Iraq finally accepted a United Nations ceasefire plan and the war ended with little gained from the original objectives.

The end of the Iran–Iraq War, however, did not mean that reports of the use of chemical warfare were to stop circulating. Within a month of the end of the war Iraq was again accused of using chemical weapons, this time against the Kurds in northern Iraq. On 25 August 1988, Iraqi armed forces began a major military offensive which caused the mass migration of more than 50,000 Kurds across the border into south-eastern Turkey. Early press reports indicated that the Kurds had fled in the face of large-scale poison gas attacks. The Iraqi government denied using chemical weapons, just as they had initially denied using them in the conflict with Iran. On 12 September, 13 countries, including Britain and the United States, asked the United Nations to investigate the allegations. The Secretary-General asked permission to send a team of experts to the region, but Iraq and Turkey denied the request. If Iraq did indeed use poison gas on Kurds living within its borders and under the rule of its government, then the events of late August 1988 raised questions about not only whether the Geneva Protocol applied but also whether a new form of human rights abuse had occurred, namely poison gas attacks by a government against its own citizens. A report by the Physicians for Human Rights (PHR), based on their mission to the region on 7–16 October 1988, revealed several conclusions for the international significance of the chemical warfare allegations.[32] First, Iraqi aircraft did indeed attack Kurdish villages with bombs containing lethal poison gas on 26 August 1988. Second, these poison gas bombs killed both humans and animals. Indeed, the report, based on interviews conducted in refugee camps, emphasised eyewitness accounts of bombing runs followed by the appearance of yellowish clouds. Survivors reported many symptoms including inflammation of the eyes and respiratory tract and blistering skin burns. The report concluded that the survivors had symptoms consistent with chemical burns by a blistering poison gas, such as mustard gas.[33] It seems impossible to determine the number of chemical munitions used and the scale of the attacks, but the pattern of injuries revealed in the PHR report were consistent with observations of victims of Iraqi chemical attacks on Iran from 1984 to 1988. Eyewitness accounts of deaths beginning within minutes of exposure, however, cannot be explained by mustard gas alone. This raises the question of whether nerve agents were also deployed. However, the absence of any real evidence, other than eyewitness reports, leaves this question unanswered.

Iraq's use of chemical weapons against the Kurds highlighted several weaknesses in the international agreements that were in place to limit the use of chemical weapons. The Geneva Protocol of 1925 was couched in broad language that left it open to abuse. Iraq clearly broke the Geneva Protocol in its war with Iran. Whether it did so in the chemical bombing of the Kurds is less clear. The Kurdish conflict fell within Iraq's border, and the protocol's referent for 'use in war' was not clear. This could have been taken to mean only in declared wars, although it could have been taken to apply to internal conflicts as well. Additionally, the Geneva Protocol only prohibited the use of chemical weapons in declared war, but modern conflicts were often not actually declared wars, even though thousands were subsequently killed.

In the summer of 1995, shortly after the fall of the United Nations 'safe area' of Srebrenica in Bosnia, survivors emerged from a long trek to safety with stories suggesting that Serb forces had attacked them during their flight with a chemical-incapacitating agent. Before the break-up of Yugoslavia in 1991, the Yugoslav People's Army's [Jugoslavenska Narodna Armija] (JNA) chemical weapons programme produced the nerve agent sarin, the blister agent sulphur mustard, the psychochemical-incapacitant BZ and the irritants CS and CN, and turned these chemical agents into weapons. In addition, the JNA also produced the choking agent phosgene, the psychochemical-incapacitant LSD-25 and the irritant chloropicrin. It is also believed they experimented with the nerve agents soman and VX, and with the blister agents nitrogen mustard and Lewisite.[34] Most of the JNA's infrastructure, production capability and expertise were inherited by the Federal Republic of Yugoslavia in 1992 and there were no indications that subsequently the stockpiles of chemical agents were destroyed. The presence of chemical agents and weapons in the former Yugoslavia was not, therefore, a secret, but this was the first allegation of their use. It was difficult to verify, but the unique character and consistency of testimonies, matched with the certain knowledge that the JNA possessed the incapacitating agent BZ and had developed both a chemical warfare doctrine and a capacity for chemical weapons use, gave strong credibility to the allegations. Furthermore, according to former JNA officers, high-ranking army officers were trained in the offensive use of chemical weapons,[35] and the JNA's manual suggested the use of a chemical incapacitant to 'create severe mental confusion'. According to the manual, a small dose of an incapacitating agent 'will affect memory, problem-solving capabilities, attention span and comprehension; larger doses will destroy completely the ability to perform any task'.[36]

The war in Bosnia began in April 1992, when the JNA, with the active assistance of the Serb paramilitary forces, instituted a campaign to deport or scare off all non-Serb inhabitants from large parts of Bosnia. Thousands of refugees fled to Srebrenica, a town that was controlled by territorial defence units loyal to the Bosnian government. Srebrenica's population surged from 8000 to an estimated 55,000–60,000 and most of these people would remain trapped in the enclave deep in Bosnian Serb-controlled territory until July 1995. Despite being declared a 'safe area' by United Nations Security Council Resolution (UNSCR) 819, Srebrenica had, in fact, become a huge isolated refugee camp. Although the United Nations was allowed in and a humanitarian crisis was thereby averted, Bosnian Serb forces occupied all of the surrounding territory and, therefore, controlled the quantity, content and frequency of deliveries to the enclave.

In early 1995, Bosnian Serb forces began planning an offensive to eliminate the three Bosnian enclaves in the territory they controlled, Srebrenica, Zepa and Gorazde. On the afternoon of 11 July 1995, Bosnian Serb forces entered the town of Srebrenica and split the population into two groups. The first group of about 25,000 mainly comprising women, children and the elderly, were bussed to Kladanj, a city in Bosnian government–controlled territory. The second group of approximately 15,000 men of military age, that is 15–60 years old, was to march to Tuzla, 50 miles to the north.

During the ensuing march a large number of people reported suffering from hallucinations. Many of the survivors concluded, either at the time or afterwards, that the Bosnian Serb forces had made use of a chemical warfare agent to disorientate the marchers and create confusion among them. In the weeks following the fall of Srebrenica, descriptions of the use of 'chemical poisons' were reported by a number of different non-governmental organisations and foreign journalists.[37] It was on the basis of these reports that the Human Rights Watch decided to undertake a preliminary investigation into the claim that Serb forces used BZ supplied by the JNA against the people fleeing Srebrenica in 1995. Some of the people who had been on the march gave testimony suggesting the Bosnian Serb forces might have used some unusual munitions. Several of the marchers gave consistent descriptions of shells that had produced a thick smoke that did not rise but spread out and of people who, following exposure to this smoke, began to act strangely and hallucinate. Most interviewed had either suffered from hallucinations themselves or observed them in others, or both. However, after interviewing survivors and United Nations personnel, and reviewing all available documentation, the Human Rights Watch found the evidence on whether a chemical agent had been used inconclusive. The evidence, whilst suggestive of the use of a BZ-like compound, was incomplete. Hard evidence, for example, in the form of chemical traces on clothes of people who died during the march, has remained elusive. However, the reason the Human Rights Watch was unable to prove the allegations could, of course, be that they were false. However, an equally plausible explanation is that the investigation in 1996 was insufficient due to two key factors: the deaths of the key witnesses and the lack of resources. The US government took the allegations seriously enough to conduct an investigation in 1997. However, the results of this investigation have not been made public, although it was suggested by the US intelligence community in late 1997 that they had uncovered information suggesting that chemical weapons were used in Srebrenica.[38]

The ban on the use of chemical weapons, as codified in the 1925 Geneva Protocol, was considered to constitute international law, applicable to all states. Yugoslavia had ratified the protocol and at no time indicated any desire to repudiate its treaty obligations. In earlier years certain countries, including the United States which used massive quantities of tear gas in the Vietnam War, maintained the protocol did not ban the use of riot control agents. However, incapacitating agents, like BZ, were not included in this apparent exception.

As discussed, the JNA was known to have been equipped with grenades and other munitions containing the hallucinogen BZ, and to have developed a doctrine for its use. Effectively used, BZ would force any persons who were hidden to betray their presence by sneezing and coughing as a result of their exposure to the smoke. They would also be likely to betray their presence by their behaviour. The full effects of BZ would start after approximately one hour and would be observable in that affected persons would start acting in an unpredictable way. Given that the column marching from Srebrenica frequently passed through woodland areas, and as such was often not visible to Bosnian Serb forces, the circumstances might have been conducive to the use of wind-borne agents that would spread out and cause disorientation among those affected by it.

Some of the testimonies collected by the Human Rights Watch concerning strange smoke and bizarre behaviour do indeed suggest that something out of the ordinary may have happened on the route from Srebrenica. However, the accuracy of the marcher's recollections must be approached with caution. Understandably, there are limits on the level of detail remembered one year after the event. After all, the primary concern for everyone on the march was to escape, not to document the experience. To this should be added the fact that the marchers had knowledge of the existence of chemical weapons in the Federal Republic of Yugoslavia. The possible expectation that the Serb forces might use chemical weapons could, then, have led to hasty conclusions. In fact, the marchers offered various theories about what had caused the hallucinations: attacks with chemical agents by Serb forces, poisoned water sources along the route, or the cumulative impact of a combination of stress and a shortage of food, water and sleep. However, one pattern emerged from the testimonies which gives credence to the chemical warfare theory. The hallucinations were predominant during the first days of the journey especially among those exposed to direct shelling attacks.[39] Ultimately, however, conclusive evidence remains intangible. Most of the people interviewed by the Human Rights Watch were in the front part of the column and therefore were not subjected to the intense bombardment suffered by those at the rear. Two-thirds of those who set out on the march from Srebrenica are still missing. Whether those who did not survive the march were victims of the use of chemical agents cannot be known with certainty from the testimonies of the survivors. There may be other types of evidence available in the former Yugoslavia. Certainly information on the development, production and stockpiling of chemical weapons is still in the possession of the United Nations Protection Force, the NATO Implementation Force and the International Police Task Force. Until these agencies are prepared to release this evidence, however, the secret of Srebrenica will never be known.

Nevertheless, it was events in Kurdistan in particular which fully illustrated both the ambiguity of what was banned and the absence of verification measures under the Geneva Protocol. Only use of chemical weapons was banned, not possession. In 1972 the United Nations General Assembly had adopted the Convention of the Prohibition of the Development, Production and Stockpiling of Biological Weapons. Chemical weapons fell outside this convention and by 1988 it became clear that a chemical weapons treaty was urgently needed to place effective constraints on the proliferation of these weapons worldwide.

Despite political efforts to abolish chemical warfare, world events dictated that chemical warfare would again be the subject of daily news reports. On 2 August 1990, Saddam Hussein sent Iraqi troops into Kuwait, allegedly in support of Kuwaiti revolutionaries who had overthrown the emirate. By 8 August, however, the pretence was dropped and Iraq announced that Kuwait had been annexed and was now part of Iraq. The United Nations Security Council called for Iraq to withdraw and subsequently embargoed virtually all trade with the country. In response to Iraq's initial invasion, on 7 August, as part of what became Operation Desert Shield, the build-up phase to the Persian Gulf War, President George Bush ordered US forces be sent to Saudi Arabia at the request of the Saudi government.

The United States' response to Iraq's invasion put the US Army's chemical warfare experience, training, production programme and lessons learned into the limelight. Not since the First World War had troops been sent to face an enemy that not only had used chemical weapons extensively within the previous few years, but had also publicly announced their intentions to use chemical weapons against the United States. Vaccines were developed and given to troops moving into the area and for nerve agent poisoning all were issued with the MARK 1 nerve agent antidote kit, which consisted of an atropine auto-injector to block the effects of nerve agent poisoning on the muscles. Pyridostigmine bromide tablets were also provided as a nerve agent pre-treatment (NAPS).[40]

The original attack on Iraq on 16 January 1991, as part of the United Nation's mandated effort to free Kuwait, designated Operation Granby in the United Kingdom and Operation Desert Storm by the United States escalated fears of a new chemical war not seen since the First World War. The initial air attack concentrated on Iraqi chemical production facilities, bunkers and lines of supply. Iraq, in response, called for terrorist attacks against the coalition and launched scud missiles attacks on Israel and Saudi Arabia in an unsuccessful attempt to widen the war and break-up the coalition. While the air attacks were going on, daily news accounts addressed the potential for chemical warfare. On 28 January, Saddam Hussein told Peter Arnett of CNN news that his scud missiles could be armed with chemical munitions.[41] US Vice President Dan Quayle, on a visit to the United Kingdom, reportedly told the prime minister that the United States had not ruled out the use of chemical or nuclear weapons and, indeed, that the United States had threatened Hussein personally if he used chemical weapons against Allied troops.[42] Nevertheless, Iraq, in turn, reportedly threatened to use chemical weapons against Allied troops if they continued high-level bombings against Iraqi troops.[43] Thus, the stage was set for what many thought was going to be the second major chemical war of the twentieth century. When the Allies began the ground war on 23 February 1991, the worst was expected and planned for by chemical defence specialists. Chemical alarms frequently went off across the battlefield, but all were dismissed as false alarms. On 27 February 1991, Allied troops liberated Kuwait City.

From the outset of the Gulf War it was known that the Iraqi regime possessed a chemical warfare capability. The large numbers of casualties that resulted from chemical weapons use during the Iran–Iraq War in the 1980s and the attack on Halabja in northern Iraq in March 1988 confirmed this. What was also clear was that Saddam Hussein was prepared to use chemical agents both in battle and against large centres of the civilian population to achieve his ends. In August 1990 the United Kingdom evaluated the capabilities of the Iraqi forces, and the initial assessment provided by Defence Intelligence Staff set out the chemical warfare agents that were thought to be definitely available to Iraq. These were the nerve agents tabun (GA) and sarin (GB), the vesicant (blister agent) sulphur mustard (H) and the riot control agent CS. Agents assessed as 'probably' available were the nerve agents Cyclosarin (GF) and VX, the vesicant nitrogen mustard (HN) and the blood agent Hydrogen Cyanide (AC). It was also known that Iraq had been provided with information on the utility and weaponisation of the nerve agent soman (GD), the choking agent phosgene, the psychochemical BZ and the vomiting agent Adamsite (DM).[44] A later assessment, in November 1990, noted that the blood agent AC was now listed as 'probably available', as was dust impregnated with sulphur mustard (H).[45] At the time of the coalition attack, Iraq was believed to have a stockpile of between 6000 and 10,000 tons of chemical warfare agents, and a capacity to produce a further 3000–5000 tons of agent per year. It was believed, however, that this stockpile included more mustard agent than nerve agent. Work carried out in Iraq after the conflict by United Nations Special Commission (UNSCOM) showed that Iraq had definitely weaponised Cyclosarin and VX at the time of the Gulf War. In addition, in February 1998, the Ministry of Defence (MOD) announced that Iraq had possessed large quantities of another chemical warfare agent, Agent 15, at the time of the Gulf War. Agent 15 was one of a large group of chemicals known as glycollates (esters of glycollic acid), usually referred to as BZ. The psychological effects of these compounds were typical of anticholinergic agents which block cholinergic nerve transmission to the central and peripheral nervous system, thereby incapacitating the victim. The immediate effects of Agent 15, had it been used, would have included dilated pupils, flushed faces, dry mouth, increased skin and body temperature, weakness, dizziness, disorientation, visual hallucinations, loss of time sense, loss of co-ordination and stupor. Many of these symptoms could be associated with, and mistaken for, influenza, but the long-term effects of exposure could certainly manifest in the kind of conditions that have been referred to as Gulf War Syndrome (GWS).

Did Iraqi forces employ chemical weapons during the 1991 conflict? The US Department of Defense (DOD) and the British MOD have long insisted they did not. Similarly, the US and British advisory committees on GWS concluded: 'Based on information compiled to date, there is no persuasive evidence of intentional Iraqi use of chemical warfare agents during the war.'[46] The absence of severe chemical injuries or fatalities among coalition forces makes it clear that no large-scale Iraqi employment of chemical weapons occurred. Nevertheless, evidence from a variety of sources suggests Iraq certainly did employ chemical weapons in the Kuwaiti theatre of operations, in other words, the area including Kuwait and Iraq south of the 31st Parallel where the ground war was fought. The *Sunday Times*, during the Gulf conflict, reported that intercepts of Iraqi military communications indicated that Saddam Hussein had authorised front-line commanders to use chemical weapons at their discretion as soon as coalition forces began their ground offensive.[47]

Other sources of evidence for sporadic Iraqi chemical warfare include, *Newsweek*,[48] US intelligence reports and military log entries which could describe the discovery by coalition units of chemical munitions in Iraqi bunkers both during and after the ground war – incidents, interestingly, in which troops reported acute symptoms of toxic chemical exposure.[49] Table 6.1 details some of the incidents in which chemical warfare agents, most notably the nerve agents tabun, sarin and Cyclosarin and the blister agents mustard and Lewisite, were detected by coalition forces during the conflict.

Although the war was a decisive military victory for the coalition, Kuwait and Iraq suffered enormous damage and Saddam Hussein was not removed from power. In fact, Hussein was free to turn his attention towards suppressing internal Shite and Kurd revolts, which the coalition did not support. Iraq agreed coalition peace terms, but every effort was made by the Iraqis to frustrate the implementation of the terms, particularly United Nations weapons inspections.

A number of reasons surfaced after the war as to why the Iraqis had not initiated large-scale chemical warfare. Some believe the remarkable speed of the coalition advance, combined with the effectiveness of the strategic bombing campaign in disrupting Iraq's military command control systems, made it difficult for Iraqi commanders to select battlefield targets for chemical attack. Furthermore, the prevailing winds, which for six months had blown from the north-west out of Iraq, shifted at the beginning of the ground war to the south-east, towards Iraqi lines.

Nevertheless, the Gulf War was, without a doubt, the most toxic war in Western military history since the First World War. Immediately after the war, allegations of chemical exposure began to surface. Initially, the British and American governments denied that any chemical exposures had taken place but veterans of the war claimed the opposite and their ailments collectively became known as GWS.

The British government's view was clearly promulgated that there was nothing unique about the illnesses exhibited by Gulf War veterans since, they argued, similar signs and symptoms had been noted among veterans from other wars, from the American Civil War, through both World Wars to Vietnam.[50] However, they were prepared in 1997 to neither confirm nor reject the possibility of GWS, but stressed further research was required.[51]

Exposure of soldiers to Pyridostigmine Bromide (PB), a toxic yellow liquid commonly used as a solvent in paints and medicines, NAPS tablets and vaccinations both in large numbers and in various types have now been recognised and acknowledged by both the British and the American governments. Additionally, in the theatre of operations,

Table 6.1 Some chemical weapons-related incidents during the Persian Gulf War

Date/location	Description	Source
19/1/91: King Khalid Military City (KKMC), Saudi Arabia	Sgt. George C. Vaughan comes under scud attack. During alert he has trouble sealing his gas mask, experiences bitter almond taste and begins choking. Within a few days he develops nausea, diarrhoea and severe fatigue. Gastro-intestinal symptoms persist after his return from Gulf, along with the development of fatty skin tumours	US House, Committee on Armed Services, Subcommittee on Military Forces and Personnel, Hearing, *Desert Storm Mystery Illness/Adequacy of Care*, 103rd Congress, 2nd Session, 15 March 1994, pp. 5–11
21/1/91: Vicinity of KKMC	French 6th Division forces report detecting two nerve agents, tabun and sarin, and mustard agent vapours	101st Airborne Division, Intelligence Spot Report, 22 January 1991
24/1/91: French base, south of KKMC	French chemical agent alarms sound after a storm blows winds out of Iraq. Chemical agent detection badges of French troops protective suits change colour, indicating the presence of nerve agent in the air	Associated Press: 'France says Gulf Troops Detected Chemicals', *The Washington Post*, 5 December 1993, p. A24
28/1/91: Saudi–Iraq border	Vulcan air defence artillery battalion detects nerve agent; detection is confirmed with M256 kit	Daily Staff Journal/Duty Officers Log, 1st Brigade, 101st Air assault, 28 January 1991
22/2/91: Saudi–Kuwait border	2nd Light Armoured Infantry, 2nd Marine Division, detects mustard agent	Central Command NBC Desk Log
24/2/91: Saudi–Kuwait border	During the crossing of an Iraqi minefield at 0635 hours a chemical reconnaissance vehicle detects mustard agent in the vicinity, at 'trace' concentrations, below an immediate threat to personnel. At 0656 hours a second FOX vehicle is sent to the minefield and confirms the presence of mustard agent. 'Unknown in origin, the chemical agent was sufficiently strong to cause blistering to the exposed arms of two crewmen'	Lt. Dennis P. Mroczkowski, *US Marines in the Persian Gulf: With the 2nd Marine Division in Desert Shield and Desert Storm* (Washington DC, 1993), p. 9

Table 6.1 (Continued)

Date/location	Description	Source
24/2/91: Saudi–Kuwait border	Marine Air Group 26 reports the detection of nerve agent at 28 degrees north, 47 degrees East on the Saudi–Kuwait border	Command Chronology, Task Force Ripper, 1st Marine Division
24/2/91: Ahmed Al Japer Airbase, Kuwait	FOX vehicle detects vapours of Lewisite blister agent. HQ responds that detection was a false alarm, caused by oil smoke. FOX vehicle separates petroleum peaks from the agent spectrum and confirms detection. 1st Marine Division Log states: 'Ripper 6 believes that chemical weapons were used, but not sure if Ripper was the target. These chemical munitions could have been exploded by our own artillery, thus causing secondary explosions'	1st Marine Division, After Action Review: Command Chronology, Testimony by gunnery Sergeant George G. Grass before Presidential Advisory Committee on Chemical and Biological Warfare Issues, Washington DC, 1 May 1996
25/2/91: Inside Kuwait	FOX team observes artillery attack to north-west at distance of about 4 km. About 5 minutes later, the mass spectrometer on the FOX vehicle sounds an alarm. The agent detected is Lewisite in a concentration to produce casualties but not death	1st Marine Division, After-Action Review
26/2/91: Inside Kuwait	Four separate reports of gas detection	Operations Log, Task Force Ripper; 1st Battalion, 7th Marine Regiment After Action Review; 8th Marines Command Chronology; 1st Marine Division, Manoeuvre Chronology
26/2/91: Inside Kuwait	3rd Battalion, 11th Marine Regiment reports that it is under chemical attack	Battle Assessment Documentation, 6th Marine Regiment

Source: Adapted from Jonathan B. Tucker, 'Iraq's use of CW in Gulf War', *Non-Proliferation Review* (Spring–Summer, 1997), pp. 116–121.

soldiers were exposed to further vaccinations, a variety of pesticides, depleted uranium (especially as ceramic dust) and chemical warfare agents. Again, these have been recognised by the British and the American governments. With regard to the administering of these vaccinations, the question that still needs to be answered is whether or not that actually caused as much harm to the soldiers as they did benefit. Studies prior to the war on the potential side effects of the NAPS and vaccinations cocktail suggest they should never have been used. According to one source, an American neurosurgeon, Thomas Tiedt, said that by giving the soldiers these 'cocktails' the military unwittingly conducted the largest clinical experiment of all time.[52] It can also be argued that the coalition effectively exposed their own troops to chemical warfare agents through the bombing of Iraqi chemical weapon production and distribution sites.

Faced with mounting evidence of chemical agent exposure, both the British and the US governments belatedly acknowledged that chemical weapons dumps were bombed by coalition troops. For the first time, the British government admitted that low levels of sarin were released after the war and that between 4 and 15 March 1991 between 20,000 and 100,000 British troops were exposed to such low levels of chemical agents.[53] Earlier in June 1997 the Pentagon had also released a report confirming that almost 100,000 US soldiers were exposed to low levels of sarin between 4 and 15 March 1991.[54] One such exposure occurred on 4 March 1991 at the Kamisiyah arsenal, north-west of Basra. After capturing the site, coalition forces blew up the Iraqi storage bunkers and, according to newspaper reports, the soldiers claimed their chemical agent detectors went off during the explosions. Later the same year a United Nations inspection team reportedly found the remains of chemical rockets and shells in one of the bunkers, and found traces of sarin and mustard agent. In 1997, the British MOD and the American DOD acknowledged that one of the bunkers probably contained sarin- and mustard-agent-filled munitions and that as many as 20,000 soldiers may have been exposed to chemical agents as a result.[55] Such incidents were described by both governments as 'unforeseen accidents'. This admission is known in intelligence parlance as a 'limited hangout'; a technique designed to show they are telling the truth.[56] They are not. Many logs relating to Gulf War experiences quickly disappeared from high security facilities in the United States, with two marines at Camp Pendleton, San Diego, publicly announcing they had observed hundreds of records from the Gulf War being destroyed.[57] As a result of such

Table 6.2 Known effects of Gulf War chemical exposures to the body

System	Vaccine	PB	OP/CB	Pyreth	Lind	NA	Mus	Du
CNS	X	X	X	X	X	X	X	X
ANS		X	X	X		X	X	
CV		X	X		X	X		
Blood	X	X	X		X	X	X	X
Immune	X	X	X					X
Muscle/bones	X	X	X	X		X		X
Respiratory	X	X	X			X	X	X
Skin	X	X	X	X		X	X	X
Renal	X	X	X		X			X
Endocrine	X	X	X		X	X		X
Genes	X		X				X	X

Source: Adapted from http://osins.sunderland.ac.uk/autism/hooper2000.htm., p. 2.
Notes: X = A known adverse effect by an agent; CNS = Central nervous system; ANS = Autonomic nervous system; CV = Cardiovascular system; PB = Pyridostigmine bromide; OP = Organophosphates; CB = Carbamates; Pyreth = Pyrethroids; Lind = Lindane; NA = Nerve agents; Mus = Mustard agents; Du = Depleted uranium.

incidents, official efforts to evaluate GWS are now irreparably flawed leading some commentators to question whether 'Pentagon Syndrome' was the real illness here.[58]

It seems clear that these exposures can be directly linked to the chronic illnesses Gulf War veterans are experiencing, the symptoms impacting on many different systems in the body causing multi-symptom, multi-organ and multi-system adverse effects. Table 6.2 examines the known agents in Iraq's arsenal and their potential effects on exposures to the body.

The Gulf War has been presented as an example of a bloodless victory by the coalition forces with very few casualties on the battlefield. The truth is rather different as Table 6.3 illustrates.

Although battlefield casualties were few, almost 15 per cent of US troops have been placed on registers of official Gulf-related illnesses. Some 26 per cent of American troops are now in receipt of benefits for Gulf War-associated illnesses. Too many post-war deaths have been reported among both British and American personnel and, indeed, in the United Kingdom one Gulf veteran has died every week since the end of the war.[59] Officially, the symptoms reported by the Gulf War veterans are described as of 'unknown origin',[60] but they nevertheless overlap significantly with the symptoms associated with Myalgic Encephalomyelitis–Chronic Fatigue Syndrome (ME–CFS) and other chronic illnesses such

Table 6.3 Numbers of deployed, dead, wounded and prisoners in Gulf War

Troops	USA	UK	Iraq
Deployed	697,000	53,000	Not known
Dead	300	49	200,000[a]
Wounded	400	Not known	300,000
Prisoner	Not known	Not known	100,000

Source: *Ibid.*, [a] BBC2 Horizon, Gulf War Jigsaw (14 May 1998). Sunderland, *ibid.*, quotes 100,000.

as Multiple Chemistry Sensitivity (MCS), which is often caused by exposure to Pyridostigmine Bromide, and even Multiple Sclerosis and AIDS (Table 6.4).

However, within the context of the Gulf War, military-level chemical exposures, it is assumed, would have been readily recognised, particularly when there was so much concern about chemical attacks. At low levels of exposure it is certainly possible that very mild cases would have gone completely unnoticed in a setting where, for example, eye irritation from sand was common, as were respiratory symptoms. It does seem unlikely, however, that typical blisters would have escaped notice, although low levels of exposure could have resembled sunburn. Additionally, in the Gulf War context, the possible agents under consideration varied greatly in the timing and onset of clinical signs and symptoms. For example, they present immediately with exposure to phosgene oxime, promptly with

Table 6.4 Common symptoms shared by chronic diseases

Symptom	GWS	ME–CFS	MCS	MS	AIDS
Joint pain	X	X	X	X	X
Fatigue	X	X	X	X	X
Headache	X	X	X	X	X
Memory problems	X	X	X	X	X
Disturbed sleep	X	X	X	Due to medicines	X
Muscle pain	X	X	X	X	X
Dizziness	X	X	X	X	X
Breathing problems	X	X	X	X	X

Source: *Ibid.*, p. 3.
GWS = Gulf War Syndrome.

Lewisite (seconds to minutes), and are delayed for hours with mustard agents. All these agents played a variety of military roles by virtue of their persistence and their ability to create vapour and contact hazards. There are those who would contend that their interaction with PB and organophosphates should be considered a possible contributory factor to GWS. Certainly, experts concur that all three, PB, organophosphates and nerve agents, when combined, can cause brain damage and/or death.[61] All could be used in conjunction with other, more toxic agents to enhance their effects, and all were dangerous as vapours, aerosols or droplets. While these agents were dangerous when ingested, it does not appear there was much opportunity for food and water contamination during the Gulf War, although there are some indications that mustard agents were disseminated and absorbed in small particles.[62] Information available is uneven with little data available on phosgene oxime, part of the nettle gas group. However, phosgene oxime is an unlikely candidate for a cause of GWS, as UNSCOM never reported this agent in the list of Iraqi chemicals they destroyed. Additionally, the chemical is so aggressive that its use would be hard to overlook.[63] There were indications of Iraqi use of an agent whose effects resembled phosgene oxime against Iran, but confirmation is lacking. Lewisite, on the other hand, although no longer considered a state of the art chemical warfare agent, remains in many countries' stockpiles. Lewisite is relatively simple and inexpensive to produce, making it attractive to nations beginning chemical warfare programmes or with limited financial resources. However, if Lewisite exposures occurred during the Gulf War, and Table 6.1 provides evidence they did, they must have been relatively mild to have escaped recognition. The setting in which FOX vehicles detected Lewisite during the conflict, near a breaching operation, was undoubtedly a possible situation for a Lewisite exposure. However, alternative explanations have been given for the readings obtained by the FOX vehicles, mainly that the road wheels, which were made of silicone can off-gas, could have resulted in false Lewisite readings.[64] Additionally, there were no reports by the United Nations that Lewisite was found in the chemicals destroyed in Iraq following the conflict.[65] Certainly no clinical cases have been associated with the reports of Lewisite detection from FOX vehicles during the Gulf War. This matter was extensively evaluated and it has been concluded that the reports were in error, reflecting only massive contamination from fire products and a misidentification of materials in the system.[66] However, it is possible that low-level exposures to Lewisite could have resembled common eye irritation and respiratory infections and such symptoms, coupled with nausea and retching, could be clue that Lewisite was the cause. However, Lewisite could not produce vomiting without eye or skin effects. There is, therefore, no conclusive evidence that coalition forces experienced a Lewisite attack or that Lewisite was in the Iraqi arsenal. The information on the long-term consequences of Lewisite exposure is not extensive. There is no indication that brief low-level exposures are associated with long-term problems, although it is recognised that Lewisite exposure can lead to Bowen's squamous cell intraepithelial cancer, a common consequence among Gulf War veterans. As shall be seen, Iraq never fully disclosed the extent of its chemical warfare programme and, indeed, consistently sought to obstruct the UNSCOM from carrying out the mandates of Security Council Resolutions (SCRs) 687 and 699.[67] Additionally, Iraq constantly sought to deceive UNSCOM about the scale of its production of the nerve agent VX and its previous use of chemical warfare agents during the Iran–Iraq War. It would, therefore, be naïve to categorically assert that Lewisite attacks did not occur during the Gulf War and Gulf War Veterans Associations should certainly persist in their demands that efforts are made to correlate epidemiological data with tactical events.

There is no 'silver bullet' to explain or cure GWS, which is not a discreet syndrome at all but a variable cluster of symptoms and disease states with different susceptibilities. The battle to cure Gulf War illnesses must be fought at the cellular, molecular and genetic levels to try to heal the delayed wounds of war and protect future soldiers. The best evidence linking toxic causes to chronic effect lies within the bodies and minds of Gulf War veterans. That evidence has been too long ignored.

Resolution 687, passed by the United Nations Security Council in the aftermath of the Gulf War on 3 April 1991, obliged Iraq to accept the destruction, removal or rendering harmless of all its chemical, biological and nuclear weapons, all ballistic missile delivery systems, and all research, development and manufacturing facilities associated with such weapons. Additionally, Iraq was ordered to submit to on-site inspections of all weapon-making facilities. To implement SCR 687, the Secretary-General was instructed to establish a UNSCOM, referred to earlier, to oversee these processes in conjunction with the International Atomic Energy Agency (IAEA). Between 1991 and 1998, UNSCOM set up an on-site monitoring and verification system, which included the installation of cameras at dual-use facilities that were capable of being converted to chemical or biological warfare purposes. By 1998, UNSCOM had eliminated 48 operational scud missiles and components, 38,000 chemical weapons munitions, 690 tons of chemical warfare agents and 3000 tons of chemical warfare precursors.[68] UNSCOM also found that, despite Iraqi claims that a project to produce VX nerve gas had been a failure, it had the capability to produce VX on a substantial scale, and had produced, between 1988 and 1998, at least 4 tons. Iraq finally admitted filling ballistic missile warheads and bombs with VX, but claimed they had been destroyed. Certainly, at the time of the Gulf War, Iraq possessed large quantities of Agent 15, sarin, tabun and mustard gas, all of which in 1991 had been deployed ready for use. In 1997, UNSCOM reported to the Security Council that it was satisfied that Iraq's nuclear weapons capability had been removed but, with regard to chemical weapons, UNSCOM had evidence that agents, their precursors and support materials for their manufacture were still unaccounted for. Although outwardly maintaining the façade of co-operation, presumably in an effort to conceal critical information about Iraq's chemical warfare programme, Iraqi officials frequently denied or substantially delayed access to chemical weapons facilities, personnel and documents. Certainly, Iraq's refusal to allow access to a range of chemical warfare sites targeted by UNSCOM contributed to the 2002/2003 crisis.

Clearly, in the absence of inspections after 1998, Iraq maintained its chemical warfare programme and, according to CIA reports in 2002, Iraq had begun renewed production of chemical warfare agents including mustard gas, sarin, Cyclosarin and VX.[69] There can be no doubt that Iraq's capability was reduced during the UNSCOM inspections and in 2002 was likely to be more limited than it had been at the time of the Gulf War. Nevertheless, Iraq's vast experience of manufacturing chemical warfare bombs, artillery rockets and projectiles suggests it is possible they possessed chemical warfare bulk fills for short-range ballistic missile warheads and a very limited number of extended-range scud missiles. Undoubtedly, all key aspects, research and development and weaponisation of Iraq's chemical warfare programme were active by 2002, and some elements were possibly larger and more advanced than they had been before the Gulf War.

So, in 2002, Iraq undoubtedly had the ability to produce chemical warfare agents within its chemical industry, although it was, in all probability, depending on external sources for precursors. Certainly, it appears that Saddam Hussein was expanding Iraq's infrastructure and, under the cover of civilian industries, was advancing the country's chemical warfare agent production capability. During the 1980s Saddam Hussein had a formidable chemical warfare capability that he used in multiple attacks against Iranians and the Kurds. By 2002, it can be assumed that Iraq probably concealed precursors, production equipment, documentation and other items necessary for continuing its chemical warfare project. As noted, Baghdad never supplied adequate evidence to support its claims that it destroyed all of its chemical warfare agents and munitions. Thousands of tons of chemical precursors and tens of thousands of unfilled munitions still remain unaccounted for. The best example is the chlorine and phenol plant at the Fallujah 11 facility. Both chemicals have legitimate civilian uses but are also the raw materials for the synthesis of precursor chemicals used to produce blister and nerve agents. Neither of these chemicals have been accounted for at that facility. Interestingly, documentation shows that Britain secretly built Fallujah 11 in 1985 and, indeed, British ministers knew all along that the 14 million plant was capable of producing mustard and nerve gas.[70]

Subsequent to George W. Bush's assumption of the American presidency in January 2001, the United States made it clear it would not accept what had become the *status quo* with respect to Iraq. As part of their campaign against the *status quo*, which from the beginning included the clear threat of the eventual use of military force against the Iraqi regime, the United States and Britain published documents and provided briefings detailing their conclusions regarding Iraq's WMD programmes and the attempts by Iraq to deceive other nations about those programmes. As a result of the American and British campaign, and after prolonged negotiations between the United States, Britain and other United Nations Security Council members, the United Nations declared that Iraq would have to accept even more intrusive inspections than the previous inspection regime. SCR 1441 was to be carried out by the United Nations Monitoring, Verification and Inspection Commission (UNMOVIC), or Iraq would 'face serious consequences'.[71] Iraq agreed to accept the United Nations decision, and inspections resumed in late November 2002. On 7 December 2002, Iraq submitted its 12,000 page dossier declaring that it had no current WMD programmes, but senior American officials quickly rejected these claims. Additionally, in his report to the United Nations Security Council on 27 January 2003, Hans Blix noted the Iraqi declaration was mainly a regurgitation of earlier documents and did not contain any new evidence to eliminate questions.[72] Consequently, over the next few months inspections continued in Iraq and the chief inspectors, Hans Blix (UNMOVIC) and Mohammed El Baradei (IAEA), provided periodic updates to the United Nations Security Council concerning the extent of Iraqi co-operation, what they had or had not discovered, and what they believed remained to be done.

Certainly, the so-called Air Force Document gave cause for concern. The document indicated that 13,000 chemical bombs were dropped by the Iraqi Air Force between 1983 and 1988. However, Iraq declared that 19,500 bombs were consumed during this period, therefore indicating a discrepancy of 6500 bombs; the amount of chemical agent in these missing bombs would have been in the region of 1000 tons. In the absence of evidence to the contrary, the United Nations Security Council assumed these quantities were unaccounted for. Similarly, at his briefing to the Security Council on 14 February 2003, Hans Blix stressed that while the inspectors had not found any WMD, they also had not been able to account for many proscribed weapons. That said, however, Blix emphasised that although the Iraqis had provided a document detailing, for example, 1000 tons of chemical agent which had not been found, this did not necessarily mean that they existed in the first place. During this same period the Bush administration in the United States and the Tony Blair administration in the United Kingdom charged that Iraq was not living up to the requirement that it fully disclose its WMD capabilities, and declared that if it continued along that path invasion would follow. On 17 March 2003, Colin Powell declared that the 'moment of truth' was arriving.[73] As Hans Blix later observed, 'The most important truth that United States spokesmen had in mind and expected to be revealed through war was undoubtedly the existence of stocks of biological and chemical weapons, and other prohibited items.'[74]

The trigger for military action was a second United Nations Resolution that would authorise an armed response. However, France, Germany and Russia argued that UNMOVIC inspections were working and should be allowed to continue. When it became apparent that the Security Council would not approve a second Resolution, the United States and Britain launched Operation Iraqi Freedom on 19 March 2003, a military campaign that quickly brought about the end of Saddam Hussein's regime. As American and British forces moved through Iraq there were initial reports that chemical weapons had been uncovered, but closer examination produced negative results. In May 2003 a specialised group, the Iraq Survey Group (ISG), comprising 1500 men, was set up to search the country for WMD. However, even before the ISG could go to work a controversy arose over the performance of American and British intelligence in collecting and evaluating information about Iraqi WMD programmes. The subsequent failure to find weapons stocks or active production lines undermined the claims by President Bush and Prime Minister Blair in their respective assessments urging war against Iraq. 'Iraq has continued its weapons of mass destruction programs in defiance of United Nations resolutions and restrictions',[75] argued the American administration. 'In recent months, I have been increasingly alarmed by the evidence from inside Iraq that . . . Saddam Hussein is continuing to develop weapons of mass destruction and with them the ability to inflict real damage on the region and the stability of the world',[76] reasoned the British administration. In addition, members of Congress and Parliament, as well as political opponents and outside observers, criticised the use of intelligence by the Bush and Blair administrations. Charges included outright distortion, selective use of intelligence and exertion of political pressure to influence the content of intelligence estimates, all in order to provide support for the decision to go to war with Iraq.

'The Iraqi Regime continues to possess and conceal some of the most lethal weapons ever devised', said President Bush, warning Iraq was intent on attacking the United States. But Mohammed El-Baradei (IAEA) concluded in March, 'There is no evidence or plausible indication of a revival of nuclear, biological or chemical weapons programmes in Iraq.'[77] The mobile chemical warfare trucks both the Americans and British warned about, also known as 'Winnebagos of Death', turned out to be mobile food inspection laboratories. Iraq's 'drones of death' that Bush warned would fly off ships to attack the United States with pestilence were, on inspection, two rickety model aeroplanes. And so it went on, a torrent of propaganda deceiving the Americans and the British into believing Iraq was armed to the teeth with WMD, was somehow responsible for 9/11, and intending, as Bush repeatedly claimed, to attack the United States and Britain.

The Butler Report, released in July 2004, strongly criticised the British government's report on Iraq's WMD. Lord Butler made it clear in the report that the claim that Iraq possessed chemical weapons ready for deployment depended on hearsay, at second-hand, from a source who turned out to be unreliable. The Butler committee concluded that they were 'struck by the relative thinness of the intelligence base . . . especially the inferential nature of much of it'.[78] Downing Street saw exactly the same intelligence and yet came to the conclusion that it 'established beyond doubt' that Iraq possessed real WMD.[79] As Hans Blix observed a week before the Butler Report, the British government refused to think critically about the evidence, even when he reported that 500 searches had produced no evidence of WMD.[80]

In 2001 the strategy of containment, in the form of sanctions and inspections, appeared to be working in Iraq and was certainly denying Saddam Hussein the ability to develop WMD. The switch from containment to invasion, it now seems clear, was not the result of any new intelligence on Iraq but due to the desire in Washington for a regime change. In their lust to invade Iraq, the Bush administration and Tony Blair deeply discredited their own nations' moral standing, credibility and democratic ideals by outrageously misleading their own people and whipping them into mass hysteria to justify a war.

7
Chemical Terrorism

We have seen terrorism emerge as one of the thorniest problems of the post-Cold War era. We have seen that terrorists are always searching for new weapons. It may not happen immediately, but somewhere, sometime in the future, terrorists will use or attempt to use... chemical weapons.

– Donna Shalala, *Emerging Diseases*[1]

The possibility that terrorists would acquire or use chemical weapons was, of course, around long before a religious cult went on the rampage in Japan in the mid-1990s. This problem had been discussed for decades, mostly behind closed doors among governments, intelligence and law enforcement institutions. A fact quietly acknowledged then, but broadcast now, is that the world is littered with facilities that contain the very materials and expertise from which chemical weapons can be manufactured. Skyscrapers, sporting arenas and transport networks have been accessible terrorist targets for decades as well. Perhaps the dilemma of terrorists obtaining and using chemical weapons did not cause undue anxiety or headlines until the 1990s because no amount of spending could alter those aspects of the threat in the past. The same is true of the present.

In 1988 the speaker of the Iranian parliament, Hashemi Rafsanjani, described chemical weapons as 'the poor man's atomic bomb'.[2] This phrase is as accurate as it is alarming. While many would argue that nuclear weapons represent the zenith of mass destruction, their construction requires advanced industrial capabilities as well as access to rare, tightly controlled materials. Chemical weapons, on the other hand, are comparatively cheap and easy to build using equipment and chemicals that are used extensively for a host of civilian purposes. With the end of the Cold War attention focused on what used to be considered 'second order' threats, and great progress was made in producing an agreement to curb the threat posed by chemical weapons. Even so, it is inevitable that the international community will continue to face threats from these weapons. Indeed, concern about potential terrorist use of chemical weapons has been mounting since 1993 and there were grave fears about possible chemical warfare activities in countries like Iran, Iraq, Libya and North Korea, to name but a few.

Terrorism is a scourge that stretches back to biblical times. Throughout history, terrorists have engaged in acts of violence to advance political causes, such as overthrowing a despot, displacing the political party in power, or trying to raise public consciousness about any number of social issues. Beginning in the 1960s, however, terrorism rose to new prominence when organisations such as the Irish Republican Army (IRA), the Baader-Meinhof Group, The Red Brigade and The Black Panthers waged well-planned, complex campaigns of violence, consisting of thousands of terrorist acts. The tools that terrorists utilised most frequently were guns, letter bombs, conventional explosives, kidnapping and hijacking, all with the objective of gathering sufficient publicity to further their aims. Throughout these campaigns terrorists were usually mindful not to kill in excess, for their success or lack thereof depended on popular support.[3] In addition to their political motives, another defining characteristic of terrorism was that its perpetrators could have been much more violent than they were but instead they observed some moral boundaries, inflicting enough violence to impact on the public, but not so much as to repulse society. 'If murder and mayhem were their primary objectives, terrorists would certainly have killed more people.'[4] Certainly, the record of terrorism indicates that most acts of terrorism during the 1960s and 1970s only involved symbolic violence. 'Terrorists want a lot of people watching, not a lot of people dead.'[5] Terrorists could rationalise their use of violence because they perceived themselves as fighters for a just cause. Terrorism, in this context, was not about killing; it was a form of psychological warfare in which the killing of a small number of people convinced the rest that they could be next. Even though terrorists in the 1960s, 1970s and 1980s carried out some shocking acts, it was possible for scholars to ascribe an element of predictability to terrorism and it was for these reasons that many experts argued that terrorists would not cross the violence threshold to WMD.[6]

However, at the end of the 1980s and as the 1990s began, the conventional wisdom that terrorists employed violence in discriminate ways was called into question as a new, more ruthless pedigree of terrorists began to leave their mark on the world. First, many terrorists that became active in this period did not embrace political causes or aim to take power. Second, a larger number of terrorists during the 1990s were intent on harming a maximum number of people.[7] Instead of kidnapping an ambassador, the 1990s style terrorists took a whole embassy hostage. In 1996, for example, members of the Tupac Amaru Revolutionary Movement disguised themselves as waiters and took 500 people hostage during a party at the Japanese Ambassador's residence in Lima, Peru.[8] Rather than hijack an aircraft, terrorists plotted to blow planes out of the sky. For example, Pan Am Flight 103 exploded over Lockerbie, Scotland, on 21 December 1988 killing all 259 people on board and 11 on the ground. An unsuccessful 1995 plot masterminded by the now convicted World Trade Center bomber, Ramzi Ahmed Yousef, sought to bring down eleven US airlines over the Pacific Ocean in a single day.[9] Additionally, during this period terrorists upped the ante from pipe bombs to truck bombs, mortars and rocket launchers, capable of blowing up entire buildings. The decade was scattered with headlines about such events. On 7 February 1991, the IRA attempted to kill the British Prime Minister John Major and his Cabinet while they debated the Gulf conflict at 10 Downing Street, by launching a mortar bomb attack. The head of Scotland Yard's Anti-Terrorist Branch, Commander Churchill-Coleman, described the attack as 'Daring, well-planned but badly executed'.[10] Certainly this attack was the boldest assault on the British mainland since the Brighton bombing in October 1984. Following the IRA attack in 1991, former British Prime Minister Margaret Thatcher, herself at the Brighton bombing in 1984, ominously stated: 'Nothing is totally safe today from a terrorist attack.'[11] In February 1993 the World Trade Center in New York was bombed by Islamic terrorists who set off a lorry bomb in the underground parking area of the centre. The attack was not overly successful from the terrorist point of view, although six people were killed. However, the bungled attack bred a degree of public complacency, as the terrorists were captured quickly and with great efficiency, but a line had been crossed: Islamic terrorists had conducted their first major operation against a target on American soil. They had failed, but it could be assumed that they would learn from their mistakes and would be back at a later date. The terrorists had actually included a container of hydrogen cyanide with the lorry bomb in the hope that the blast would drive gas up the ventilation system of the building, but the gas was incinerated in the explosion. However, it was a chemical weapons attack, and therefore, arguably, another line had been crossed. Such events continued into the decade, the bombing of Khobar Towers in Saudi Arabia in 1997 and the US Embassies in Kenya and Tanzania in 1998 to name just a couple of examples.

A common thread through much of late twentieth-century terrorist activity was religion. Religious wars, long ago, became a staple of history, and it has been commonplace as well for charismatic leaders to sweep thousands into religious cults, only to lead their followers seriously astray. Jim James, for example, persuaded 900 people at his Guyana mission to swallow Kool Aid poisoned with a cyanide cocktail in 1978. Marshall Applewhite of Heaven's Gate sent 39 to their death hoping to ride a spaceship they believed was hiding behind the Hale-Bopp Comet in 1997, as a result of drinking a cocktail laced with Phenobarbital.[12]

More worrying, what also happened in the 1990s in the world of religion and cults, was that some groups bound by religion began to acquire weapons, including the Islamic groups Hezbollah and Hamas. On the surface, religion and a willingness to engage in wholesale killing appear incompatible, but there is strong evidence that religion was a causal factor in the 1990s surge in the level of terrorist violence. Although religious terrorists were responsible for only 25 per cent of the terrorist acts in 1995, they caused 58 per cent of all fatalities attributed to terrorism that year.[13] This causal link is easily explained when it is recognised that some religious terrorists worship a god that says it is permissible to kill indiscriminately. Then, the constraints of morality fall away since religious terrorists see violence as a sacramental act or divine duty. In this way religion serves as the legitimising force, specifically sanctioning wide-scale violence against an almost open-ended category of opponents.

Religion, however, was not the only driving force behind the rise in terrorist violence. The far Right of the political spectrum which invokes nationalism, individualism, ethnic or racial belligerence and capitalism, also exhibited a predilection for large-scale violence. Extremist groups circled the globe, each rallying around an ideology tailored to their particular circumstances, from the British National Party (BNP) and neo-Nazis to militias and animal action groups. Even environmentalists, who have a pacifist, tree-hugger image, sometimes fall into the extremist category.

Several other reasons help to explain the growing terrorist carnage. First, the public generally became hardened to 'ordinary' violence and so terrorists became encouraged to stage ever more sensational attacks to grab the media spotlight. Second, terrorists became masters of their craft, acquiring smaller and more effective weapons, flexing their creative muscles and getting better at evading capture. Also, some terrorist organisations became beneficiaries of an influx of resources from organised crime, amplifying the destructive capability of groups that otherwise might not have been as effective or indeed active. Finally, some terrorists no longer felt the need to claim credit for their handiwork and the veil of anonymity liberated such individuals to engage in ever more deadly tasks.

In contrast to nuclear or biological terrorism which remain problematic and complex matters to carry out, and in the case of nuclear terrorism still largely theoretical, chemical terrorism is more concrete and practical and in several instances has already been used. The most conspicuous chemical terrorist attack occurred in early 1995, when members of the Aum Shinrikyo (Supreme Truth) cult in Japan released toxic gases at various targets, particularly the underground systems of Tokyo and Yokohama, injuring hundreds and killing several. Fortunately the number of victims did not reach higher proportions, despite the high toxicity of the material released and the panic that understandably gripped the underground passengers and caused a stampede from the sites involved. This was the first time an extremist organisation had attempted to use a chemical substance in a mass terrorist attack. It was not, however, the first use of chemical agents by terrorist organisations in order to instil terror, carry out blackmail or cause large-scale economic damage to their rivals. In several cases, various organisations in different parts of the world have laced food products in order to sabotage the marketing of them and terrorise consumers. This was the case with citrus fruit exported from Israel to the European markets in which a chemical substance was injected in order to cause economic damage to Israel.[14] In another case a chemical agent was injected into chocolate confectionery in Japan to blackmail the manufacturers,[15] and in 1991 a chemical toxin was found in jars of baby food in the United Kingdom and the United States.[16]

Terrorists packing guns and bombs are frightening enough, but chills go down the spine at the thought of terrorist organisations employing weapons that cannot be heard, seen, smelled or tasted. The shocking attack by Aum Shinrikyo was proclaimed by many as the dawn of a new age catastrophic terrorism involving chemical and nuclear weapons.[17] Other observers believed it was an aberration, an isolated incident that was unlikely to be repeated. Still others believed it illustrated a fundamental change in the proliferation threat, illustrating the ease with which terrorist groups could acquire and deploy chemical weapons capabilities undetected.

Before the underground attack most Japanese citizens saw Aum Shinrikyo as an oddity, but few perceived them as a menace to society. The Japanese are renowned as a people of religious tolerance and were accustomed to sects advocating various paths to enlightenment and salvation. Aum Shinrikyo's businesses revolving around herbal teas and medications, health clubs and computer-related enterprises, were also within the mainstream. It seems the only ones who suspected something ominous were the citizens who lived near the cult's compounds and, belatedly, Japanese law enforcement authorities who slowly began to connect Aum Shinrikyo to such crimes as kidnapping and murder.

Despite an almost 'concentration camp' environment,[18] Aum Shinrikyo's followers numbered tens of thousands and financially its investments were somewhere between £2150 million and half-a-billion pounds.[19] Before long Aum Shinrikyo's claws reached overseas, particularly to Russia where the cult had several branches in Moscow and 11 outside the capital city. Not coincidentally, several of these branches were located near important missile, chemical and biological weapons facilities. Throughout the world a premium was placed on the recruitment of scientists and technicians to aid the cult's armament programmes. Generally, the cult set its sights on futuristic weapons, like lasers and seismological weapons, but it also sought to harness older weapons, disease and chemicals.

Unable to buy chemical weapons, Aum Shinrikyo initiated its own poison gas programme in the spring of 1993.[20] Stocked with ingredients, Aum Shinrikyo's scientists then busied themselves in the cult's high-technology laboratories exploring and producing chemical agents. VX was synthesised on five occasions and, additionally, the cult's chemists reportedly knew their way around tabun, soman, mustard gas, hydrogen cyanide and phosgene. Nevertheless, the cult did not produce these agents in quantities, for example, only one pound of mustard gas was manufactured.[21] Aum Shinrikyo produced their first batch of the agent that was to make them infamous in the autumn of 1993 when they synthesised 20 gm of sarin.[22] However, because the cult's biological concoctions were not achieving the desired results, Aum Shinrikyo's chemists took steps to prove that its sarin would kill. Tests on live subjects took place at a ranch purchased in the Australian outback by the cult in September 1993. When Australian police later investigated the ranch they found the remains of 29 sheep. Later experiments conducted on the wool and soil samples revealed trace residues of sarin.[23] It seems, therefore, that by gassing the sheep, the cult had obtained unmistakable proof of their sarin's lethality.

On 27 June 1994 a truck left the Aum Shinrikyo compound for the city of Matsumoto. At 2200 hours the truck, customised to disperse poison gas, pulled into a supermarket car park and at 2240 hours started to spray 20 kg of liquid sarin into the night air, leaving the wind to disperse the agent. The evening was warm and seven residents had gone to bed leaving their windows open. These people, who came down with a runny nose, cough and shortness of breath, may have thought they had the onset of a cold; in fact they were suffering from the first symptoms of sarin exposure. Within 1 hour, 253 people attended outpatient departments, 58 were admitted to hospital and 7 lost their lives. Only later would laboratory analyses reveal the culprit to be sarin. The attack was originally dismissed by the authorities as an accident, caused by a chemical hobbyist who was tinkering with pesticides. That rationalisation was implausible, but the idea that somebody wanted to kill large numbers of Japanese citizens at random with nerve gases was even harder to believe. In September 1994 an anonymous paper on the Matsumoto sarin attack was sent to Japan's major media outlets. The document asserted that Aum Shinrikyo was responsible for the attack and that Matsumoto was an open air 'experiment of sorts', and noted that the results would be much worse if sarin were released indoors, for example, in a 'crowded subway'.[24] Certainly the cult had a surprisingly well-developed technical infrastructure which included front companies for purchasing materials and equipment, well-equipped laboratories, extensive chemical manufacturing facilities and several 100 tons of 40 different kinds of chemicals. One estimate suggested that the materials together could have produced about 50 tons of chemical weapons agents, enough to kill as many as 4.2 million people.[25] Certainly Aum Shinrikyo had an enormous budget, spending £16 million on its poison gas plant alone.[26] However, it should be noted, it would be possible to produce lethal chemical weapons in sufficient quantities for use in terrorist attacks with far more modest resources than those of the Japanese sect.

Among the Monday morning commuters on 20 March 1995 were five Aum Shinrikyo terrorists preparing to board trains at previously assigned times. One carried three sarin filled packages and the other four carried two each. Between 0746 and 0801 hours the terrorists stabbed the plastic liners of the packages with umbrella tips and then calmly stepped off their trains several stations later where they were met by their getaway drivers.[27] Meanwhile, the trains continued to the centre of Tokyo, and the packages that had been left on the floor began to seep sarin vapour and liquid agent. Of the eleven bags, eight were ruptured and three were recovered later intact. Police estimated that a total of 159 ounces of sarin were released on the five trains hit.[28]

Some commuters did not notice a smell before the fumes began to choke them, others described odours like burning rubber or mustard and a 'gooey' substance on the floor. Nevertheless, the effects of the sarin were pronounced as passengers began coughing, collapsing, vomiting and convulsing within a few stops from where the terrorists disembarked and where the sarin had been released. The death toll on 20 March was 12.[29] The after effects of the attack, however, were felt by far more than just those unfortunate enough to be on the underground system during the morning rush hour. Prior to this attack, Japan had enjoyed a reputation for being one of the safest, least violent countries in the world. Aum Shinrikyo's attack shook the entire nation in a way no earthquake ever had. Also, the attack was an alert to authorities around the world and it is clear from the testimony from the leader of Aum Shinrikyo, on his capture, that other targets of the sect included Washington, New York and London.[30] Aum Shinrikyo, it is fair to say, changed the way the world thought about terrorism, but by almost any standard, they were a terrorist nightmare – a cult with money and technical skills led by a man with an apocalyptic vision, an obsession with chemical weapons and no qualms about killing. Nevertheless, talk of 'unconventional' terrorism moved from academic and government circles to Hollywood, which was unable to resist a story line as delectable as terrorists menacing the United States with chemical weapons.[31] Novelists found the subject just as enticing and it was the novel *The Cobra Event* that caught the eye of the former US President Clinton prompting him to tell *The New York Times* he was so worried about unconventional terrorism that it caused him sleepless nights. He commented: 'I would say that it [unconventional terrorism] is highly likely to happen . . . in the next few years.'[32] Taken together, films, novels and the president's interview meant that by the late 1990s the world was awash with elected officials issuing warnings about the perceived chemical threat. Such statements conveyed an impending doom that left some people perplexed. What had changed so dramatically overnight to warrant such alarm about unconventional terrorism? Terrorism had been around for centuries, and while there were lessons to be learned from Aum Shinrikyo's attack, it was somewhat rash to predict that terrorists henceforth would embrace chemical weapons, especially since the regular tools of their trade were much easier to acquire and use and, importantly, served their purposes equally well. Certainly saturated media coverage took its toll on, in particular the American psyche. According to a survey published in 1999, 84 per cent of Americans viewed international terrorism as the most serious threat facing their country. Chemical weapons were the second most feared threat, according to 76 per cent of those questioned.[33] On 11 September 2001 the issue of a major terrorist attack on the United States ceased to be theoretical. Radical Islamic terrorists of the Al-Qaeda group hijacked four airliners on domestic US flights, took over the controls and flew two of them into the twin towers of the World Trade Center in New York City and one into the Pentagon. The other crashed in a field after the terrorists fought with passengers. Total casualties from the operation, which was as meticulously planned and executed as it was ruthless, were over three thousand people. No chemical agents were used on this occasion, but this should not be taken to rule out terrorist use in the future. Certainly one lesson from Tokyo was that it presaged a new age of terrorism wherein chemical weapons would be an integral part of the terrorist arsenal.

As one analyst observed: 'Television interviews where public officials describe how many people a drop of VX will kill terrify, but do they enlighten?'[34] It is true that major features of modern society, skyscrapers, sporting arenas, amusement parks and transport networks are, by their nature, open to terrorist attack. Indeed, Islamic terrorists who have been captured and interrogated have said that their training in Osama Bin Laden's Al-Qaeda camps in Afghanistan involved exercises in dispersing hydrogen cyanide into the ventilation systems of buildings, which could potentially be an effective and lethal tactic.[35] There is also the possibility that terrorists could target a plant that manufactures dangerous chemicals and sabotage it to spread a huge toxic cloud, much like the 1984 accident at the Union Carbide plant (owned by Dow chemical, USA) in Bhopal, India, which killed over 2000 people. Moreover, the chemical weapons options available to terrorists appear abundant. However, when scary 'what-if' scenarios, almost to the exclusion of technical and historical analysis, dominate the problem of chemical terrorism, pragmatic discussions are lost in the 'dense smog of terrorism'.[36] It is clear that those who speak so alarmingly about a chemical terrorism threat are not aware that in the twentieth century not a single European or American died as a result of bio-terrorism and only one US citizen died in a terrorist attack involving a chemical agent.[37] Do those who opine about the chemical terrorist threat know or appreciate that amassing from scratch a genuine mass casualty capability with poison gas is not exactly easy? But, it is not impossible. For the organisations involved, chemical terrorism has several clear-cut advantages over conventional terrorism. Chemical substances are easily accessible and available, they can be manufactured using simple chemical processes known to any university student and the components are usually simple products that can be obtained in the free market without restrictions. A chemical attack then can be executed using, for example, off-the-shelf pesticides sold in most supermarkets. If this were not enough, many countries, including Third World countries and countries that are known supporters of terrorist organisations, have large arsenals of chemical materials. Therefore, it appears quite likely that there is a strong possibility that chemical weapons could be transferred from one of these countries – particularly from those who have not flinched from using chemical weapons against their own citizens and in their wars with neighbouring countries – to a terrorist organisation in order to execute attacks. In October 1998 it was finally confirmed that an El Al Boeing 747 cargo aircraft that crashed near Schipol Airport in Amsterdam in October 1992 had been carrying a shipment of DMMP, a dual-use chemical used as an ingredient in the manufacture of nerve gas, destined for Israel's chemical warfare unit in Ness Ziona.[38] In the months and years after the El Al crash, hundreds of people living near the crash site along with the rescue workers developed inexplicable illnesses ranging from breathing problems to skin rashes, nervous disorders and cancer. It was suspected that the illnesses stemmed from exposure to toxic chemicals carried by the Israeli aircraft, which burned after the crash. The confirmation that DMMP was on board was provided only after a full six years in which Israel had refused to provide a full accounting of what the plane was carrying; Israel admitting only in the end that it was a 'commercial cargo'. The contents of the cargo, some 20 tons, apparently shipped by the Israeli Defence Ministry, have yet to be fully identified. El Al's lawyer, Robert Polak, told the Dutch government that the details would never be forthcoming because of what he termed 'state security reasons'.[39]

Chemical terrorism is relatively inexpensive and it does not require extensive facilities. Toxic chemical substances are cheap to purchase and manufacture (at least, when compared with, say, nuclear substances) and so resource-starved organisations can obtain and use them very easily. Chemical substances also have the advantage of mobility. In contrast to a nuclear bomb, which is usually large and cumbersome and requires special vehicles and security in transport, small amounts of chemicals can be easily and covertly transported – and only a small amount of a chemical is needed for an attack. Over the past 20 years modern technology has made great progress in the detection of conventional weapons and explosives, and many countries have put this technology to practical use at sensitive facilities and in congested population areas. The technology for the detection of chemical substances, in contrast, has not been disseminated widely, and has not been used for preventative security in the foiling of terrorist attacks. This, of course, facilitates the terrorist's work in delivering chemical substances to their targets and concealing them there with very little chance of prior detection.

Aum Shrinrikyo's sarin nerve gas attack against Tokyo commuters led to a significant, and understandable, concern that terrorists might, from now onwards, use unconventional weapons. Certainly, if a tactic attracts widespread attention, other terrorists will usually want to imitate it.[40] Three questions immediately present themselves then. Why would terrorists replicate Aum Shinrikyo's attack? Who, among the spectrum of terrorist groups, would be the most likely to copy such an attack? Why has the Aum Shinrikyo attack not been duplicated?

A number of indications suggest that the new generation of terrorists are nihilistic, apocalyptic and wrathful – in short, groups which would consider using any instrument of violence. Unlike customary attacks using guns and bombs, a gas attack is a guaranteed front-page headline with a significant psychological impact. According to one definition of a successful terrorist attack, Aum Shinrikyo's sarin assault was a resounding success. That definition revolves around the amount of publicity that a terrorist operation gains.[41] Few would quibble that Aum Shinrikyo's attack was one of the most publicised terrorist incidents in history prior to 9/11. Terrorists may resort to chemical weapons for several reasons: to kill as many people as possible, to incite widespread panic, to establish a position of strength from which to negotiate their demands, to enhance their ability to execute attacks anonymously, or to disrupt and significantly damage a society or an economy. Certainly the first and third of these points came to the fore in terrorist attacks in the late 1990s and early 2000s. Additionally, and of course, perhaps more disquieting, terrorists might select a chemical agent attack to fulfil a biblical prophecy.

Unsurprisingly, given recent terrorist trends, numerous experts contend that religious or apocalyptic groups head the list of terrorists thought to be most likely to employ chemical weapons. According to Bruce Hoffman, the conventional wisdom that terrorists do not know how to work with chemical weapons, that they will be held back by moral, political or operational impediments, or that they do not want to kill large numbers of people is 'dangerously anachronistic'. He asserts that the combination of religious terrorism and chemical weapons 'could portend an even bloodier and destructive era of violence'.[42] Jeffrey Simon argues that religious and apocalyptic groups exhibit at least two important characteristics that might drive such a group to obtain and use chemical weapons, namely that they are not concerned that potential supporters might react negatively, and that they have a previous track record of large-scale violence.[43] Other categories of terrorists that could explore the chemical weapons option include those who might obtain state sponsorship, although this would be unlikely since state sponsors would recognise that if they were identified as the source of a mass casualty weapon used by terrorists, the retaliation would be severe. Also to be considered are extreme single-issue groups, fanatical nationalists, right-wing militias and terrorists desperate to make their point. A final category of terrorist that might be motivated to use a chemical agent is the psychopath.[44] Nevertheless, reassuringly, this final category of terrorist – the unaffiliated madman – is least likely to have the financial and technical wherewithal to mount a successful attack causing mass casualties.[45]

Although the newspapers, airwaves and bookshelves were full of catastrophic predictions from the late 1990s onwards, there are several reasons why terrorists have refrained from, and may well continue to avoid, the deployment of chemical agents. Some terrorists might have moral objections to the potential scale and appalling nature of the death that such weapons can cause. Since so many could feasibly die in a mass attack, it is clear that extravagant demands would be made by the terrorists. If terrorists use chemical weapons but do not ask for much, the public would urge the government to grant the demands. The larger the number of possible victims, the more vocal the public is likely to be. Some terrorist groups, on the other hand, might be concerned that the use of chemical agents might anger potential supporters and possibly offend group members. Finally, governments are likely to react with vehemence should terrorists release chemical agents. When combined with the fact that terrorists can still attempt to achieve their goals using conventional weapons, these reasons make chemical attacks unattractive options. Terrorists must also have considered whether Aum Shinrikyo's use of chemical weapons helped them achieve their goals and, indeed, if anything, the underground attack backfired. The Japanese government remained intact and most of the cult's hierarchy was imprisoned, some under death sentences. The cult itself persists but is a shadow of its former self. In this light Aum Shinrikyo's venture into chemical warfare can hardly be viewed as contributing to the cult's overall objectives. Accordingly, it could be argued that terrorists see Aum Shinrikyo's sarin attacks less as a beacon but more as a warning not to follow suit. Taking this line of argument to its natural conclusion, terrorists then will shun chemical weapons in favour of their familiar, less complicated trade tools, guns and bombs. Table 7.1 shows that, indeed, since the 1995 Tokyo attacks terrorist attacks using conventional methods have increased worldwide.

Other than motivation and objective, the major question that hangs over the subject of chemical terrorism is whether the terrorists have the

Table 7.1 International terrorist attacks (by type) 1996–2000

Type	1996	1997	1998	1999	2000
Firebombing	1	0	5	12	20
Armed attack	3	5	5	11	6
Hostage	6	8	4	21	40
Bombing	55	108	96	111	251
Hijacking	0	0	0	3	2

Source: (1996–1999), *Patterns of Global Terrorism 1999*, Washington DC: US government Printing Office (April 2000), www.terrorthreat/map/s/orgn.htm.

technical and operational wherewithal to acquire and use this type of weapon effectively.

Chemical weapons are highly toxic liquid and gaseous substances that can be dispersed in bombs, rockets, missiles, artillery, mines and grenades or spray tanks. If absorbed through the skin, these man-made substances can incapacitate or kill with just a few microscopic drops. Most chemical agents dissipate quickly when released into the atmosphere, while a few, notably VX and mustard gas, are far more persistent and raise long-term health and environmental concerns. Since chemicals are integral to countless consumer products and the chemical industry can be found in every corner of the world, the chemicals and equipment used to produce for the domestic market can easily be used to produce chemical warfare agents. Of particular relevance to poison gas and nerve gas production are the chemicals used to make pesticides, fertilisers and pharmaceuticals. These chemicals can be purchased on the open market from commercial traders, laboratory suppliers, and even by mail order and on the Internet. Table 7.2 illustrates the dual-purpose nature of chemicals, showing the commercial uses for the chemical ingredients of the principle chemical warfare agents.

Through an organisation known as the Australian Group, set up in 1985 following the use of chemical weapons in the Iran–Iraq War, 30 nations enforced uniform export controls on chemical precursors and manufacturing equipment to attempt to thwart the progress of terrorist organisations. If a member country denies an export license out of concern for possible weapons use, the rest of the Australia Group is informed of the decision and the items requested. Such sharing of information enables these 30 nations to work together.[46] For domestic sales, however, these licensing procedures do not apply. In theory, therefore, a terrorist organisation could purchase average quantities of the requisite chemicals

Table 7.2 Commercial application of chemical weapons precursors

Type of chemical agent	Commercial uses of the chemical ingredients
Mustard agents	Lubricant additives; ballpoint pen ink; manufacture of plastics, paper and rubber; photographic developing solutions, textile dyes, pesticides, chlorinating agents, engineering plastics, cosmetics, detergents, pharmaceuticals, insecticides, waxes and polishes, toiletries, cement additives, resins.
Tabun	Petrol additives, hydraulic fluids, insecticides, flame retardants, pharmaceuticals, detergents, pesticides, missile fuels, vulcanisation of rubber, extraction of gold and silver from ores.
Sarin	Flame retardants, petrol additives, paint solvents, ceramics, antiseptics.
Soman	Lubricants, cleaning and disinfectants for brewery, dairy and other food processing equipment.
VX	Organic synthesis, insecticides, lubricant oil, pyrotechnics.

Source: Adapted from CIA, *The Chemical and Biological Weapons Threat*, Washington DC (March 1996), pp. 9–16.

from a national firm or even local shops without arousing suspicion. For that matter, a few of the components of some of the most dangerous chemical weapons can be found beneath kitchen sinks and in garden sheds.

As with chemical weapons ingredients, the chemical equipment needed to make chemical warfare agents is commercially available just about anywhere. Certainly, to set up a full-scale poison gas production line, terrorists would need reactors and agitators, chemical storage tanks, containers, receivers, condensers for temperature control, distillation columns to separate chemical compounds, valves and pumps to move chemicals between reactors and other containers. Additionally, ideally the equipment would be corrosion-resistant. For a full-scale mustard gas production plant the price tag would be between £2.5 and 5 million. Approximately £10 million would be required to set up a plant to manufacture tabun, sarin or soman.[47] Terrorists, however, can be assumed to forego the scale and the safety precautions that most governments would consider essential for a weapons programme. In fact, standard process equipment or a laboratory set-up of beakers and tubes would be adequate to make smaller quantities of poison gas. It is therefore possible that chemical warfare agents could be made in a kitchen in quantities sufficient for mass casualty attacks, at a fraction of the cost of setting up chemical plant.

According to some accounts, any individual with A-level chemistry has enough knowledge to make poison gas. However, all chemical reactions are not alike and A-level chemistry students rarely work with the extremely toxic chemicals involved in the synthesis of chemical warfare agents. Expert opinion, therefore, weighs heavily in favour of graduate-level chemical skills being a prerequisite for chemical weapons work. Of course, British, American and Russian universities turn out hundreds of such post-graduates each year and such individuals know exactly where to locate the formulas for chemical warfare agents. Mustard agents came of age during the First World War and nerve agents were discovered in the 1930s; the production processes used then are still viable today. Indeed, few military technologies have changed *so little* as chemical weapons over the past century. The synthesis of the various agents has been a frequent topic of books, professional papers and patents. Indeed, as early as 1970 the open literature already contained 15 formulas for the 'V' class of nerve agents. Which chemical agents terrorists might choose is, however, an issue under much debate. Some argue terrorists would bypass choking agents because they would have to make such large quantities for an attack, along with blister agents because they would only harm, and not kill, large numbers of people. This school of thought believes terrorists would favour sarin, a highly toxic nerve gas that is relatively easy to manufacture. Others argue that although large quantities of mustard gas would be necessary for an attack, mustard gas is easier to manufacture than nerve agents.[48] Finally, the CIA in the United States classified choking, blood and blister agents as 'relatively easy to produce' and nerve agents as 'somewhat more difficult to produce'.[49] Ultimately, then, it seems the manufacture of both mustard and nerve agents are within a determined terrorist's ability.

The options for delivering chemical agents are varied. Toxic chemicals could be employed to poison food and water supplies, put into munitions or distributed by aerosol. Chemical agents could also be the payloads of any conventional munition, from bombs and grenades, to artillery shells and mines.[50] Sprayers can be mounted on aircraft, lorries or boats, and crop dusters could be employed.[51] Particularly when dispersing a chemical agent outdoors, it must be assumed that 90 per cent of the agent will not reach the intended target in doses sufficient to cause large casualties because chemical agents are susceptible to weather conditions. Given the operational challenges of dissemination, while a single person might be able to execute a chemical attack in a building against a group of people, it is doubtful that such an attack could produce more than a few hundred casualties.

Nevertheless, it is important in assessing the chemical terrorism threat to consider what sorts of weapons, in reality, are available to terrorist organisations. The most common agents are sarin, tabun, hydrogen cyanide and VX. Since tabun is so easy to make, it is probable that any potential chemical terrorist would have access to it. During the United Nations fact-finding mission to Iraq, officials found several packets of atropine, the antidote to VX, and this in turn begs the question of why Iraq would have the antidote if they did not also have the chemical. Possible answers to this question include that Iraq did, in fact, have a stockpile of VX (although no evidence has ever been produced that they did) or they had close ties to a group that did. Alternatively, they may have feared an American VX attack in the future. Sarin, soman and Lewisite could also pose a possible risk as they once comprised the Soviet Union's stockpile. Since the former Soviet Union had a long (rumoured) history of selling weaponry to terrorist groups, it is not too large a mental leap to assume they could have sold their chemical agents to a potentially volatile organisation. However, just because terrorist groups have access to deadly chemicals, it does not necessarily mean they will use them. The purpose of terrorism is, after all, to cause fear in order to influence; the threat of VX would cause fear without the chemical actually being deployed.

Historical analysis can provide a much sounder understanding of what terrorists have and have not done with chemical agents. The RAND Corporation database kept since 1968, which tallies international terrorist events, shows over 9000 terrorist incidents up to 2003.[52] Terrorists have been involved with WMD on less than 100 occasions, inclusive of attempts to acquire or make their own devices, threats to use such weapons, or cases of actual uses.[53] Conversely, according to Amy E. Smithson, who uses statistics from the Monterey Institute of International Affairs,[54] in the period 1975–2000 there were 139 cases in the United States alone related to politically or ideologically motivated groups who were in some way connected to chemical substances.[55] Internationally the figure was much higher, standing at 203 cases.

Certainly, particular terrorist groups are far more likely to get involved with chemical agents than others are. Taking international cases, which total 342 for the period, unsurprisingly 18 per cent are attributed to fundamentalist religious organisations, but a surprising 23 per cent are connected to nationalist groups.[56] Insight into the gravity of the chemical terrorism threat can also be gained by examining the frequency of terrorist activities. For analytical purposes a line must be drawn between activities of lesser and greater concern. Sometimes talk of a plot is idle, but accordingly when terrorists move past plots to acquisition, possession and/or use, a more serious and purposeful terrorist association is signified. Table 7.3 shows that the most frequent activity internationally is in cases of use.

The range of terrorist activities connected with chemical agents can be further divided into those justifying possible serious and possible grave concern. In Table 7.3, point 2 categories could be reasonably placed as causing serious concern, while point 3 categories undoubtedly signify a grave concern. Should the bulk of these cases involve extremely lethal substances, the data would lend belief to forecasts that terrorists are leaning towards an escalation involving mass casualty attacks with poison gas. Conversely, if the chemical substances involved are incapacitating as opposed to lethal agents, then the data would support the argument that terrorists are neither inclined nor capable of launching mass casualty attacks with these weapons. In the years surveyed by the Monterey database,[57] terrorists attempted to acquire, possess (or possessed) and threatened to use over 60 types of chemical agent. Various types of cyanide accounted for 45 cases, sarin for 9 cases and insecticides and pesticides for 7 cases. The nerve agent VX was involved in 6 cases. A few important observations can be made about the items listed. Certainly these substances have ample killing potential and some of the substances are known to have been weaponised by governments. The three substances that terrorists used most commonly were butyric acid (22 cases), cyanide (20 cases) and tear gas (14 cases). Of course, cyanide immediately draws attention because of its lethality and battlefield use

Table 7.3 Terrorist activities with chemical agents (1975–2000)

	Type of activity	No. of incidents
1.	Hoax	37
	Plot only	19
2.	Attempted acquisition	8
	Possession	42
	Threat with possession	7
3.	Use of agent	90

Source: Adapted from Amy E. Smithson, *Ataxia: The Chemical and Biological Terrorism Threat and the US Response*, Washington DC: Henry L. Stimson Center (2003), p. 61.

during the First World War. The Monterey database attributes 12 deaths and 63 injuries to terrorist use of cyanide.[58] The foul-smelling butyric acid was deployed most notoriously in attacks on abortion clinics in the United States in the summer of 1998.[59] Tear gas is a well-known incapacitating agent used by law enforcement agencies for riot and crowd control purposes throughout the world. In all, the Monterey database records 154 fatalities due to chemical terrorism in a 25-year period. Another clue then to the absence of more casualty cases may lie in the terrorist choice of delivery methods, all of which can be categorised as low technology, as Table 7.4 illustrates.

What the data in Table 7.4 does not reveal is whether the terrorists selected these delivery methods because they could not do better or because their intended targets were an individual or small group, as opposed to a large crowd.

The evidence presented demonstrates that terrorists certainly have acquired and threatened to use a larger array of chemical agents than they have actually deployed. Indeed, the statistics charting terrorist behaviour with chemical substances from 1975 to 2000 show that by far the most frequent of terrorist activities were hoaxes, which are a poor indicator of true terrorist intent to pursue such capabilities and to use such weapons. What does not become clear from the evidence, however, are the reasons terrorists did not cross the threshold to use, although this could simply come down to individual circumstances. The inhibiting factors could range from interruption by military or law enforcement agencies prior to action, moral qualms, fear for personal safety or, simply, the inability to overcome the technical challenges of producing and

Table 7.4   Methods employed by terrorists to deliver chemical agents

Delivery method	No. of events
Direct contact	33
Aerosol/spray	21
Food/drink	13
Consumer product tampering	10
Explosive device	6
Water supply	5
Canister	1
Letter/parcel	4
Injection/projectile (bullet)	1

Source: Adapted from Amy E. Smithson, *Ataxia*, p. 64.

dispersing chemical agents. In the final analysis, however, it is clear that conventional terrorism is far more widespread, far more harmful and far more deadly than chemical terrorism. Therefore, if the past is any predictor of the future, terrorist incidents involving chemical substances will continue to be small in scale and far less harmful than conventional terrorist attacks.

8
Controlling Chemical Weapons

The entry into force of the 1993 CWC on 29 April 1997 was unique in the history of arms control. This agreement both banned an entire class of weapons and simultaneously addressed chemical proliferation concerns. It was not, however, the attention to non-proliferation that made the Convention unique, rather that the CWC incorporated an elaborate international system for verification of compliance.[1]

Chemical warfare, as it is understood today – the military exploitation of the properties of certain chemicals against people or the environment – began in April 1915. The use of chemical weapons since have been confirmed by Italy in Abyssinia, Japan and China in the 1930s and early 1940s, by Egypt in the Yemen, by Iraq against Iran, and in Korea, Vietnam and the Gulf War. In March 1995 the first major terrorist incident using chemical warfare occurred when the nerve agent sarin was released in the Tokyo underground system.

Certainly, some states in unstable regions of the world remain interested in chemical weapons. While they are probably not the most effective battlefield weapons, chemical weapons can instil terror in entire populations and compel governments to strike pre-emptively against chemical weapons production and storage sites. The presumption of a chemical warfare capability in an adversary state can magnify an existing condition of crisis instability. The CWC offered the prospect that in the not too distant future an entire class of unconventional weaponry would be eliminated.

Attempts to ban chemical weapons progressed from early restrictions on their use to their total prohibition and elimination. Abhorrence against the use of poison in war can be found in some of the oldest literature of several cultures.[2] In the nineteenth century the international community began to codify the conduct and customs of war, which resulted in the

1899 and 1907 Hague Regulations Respecting the Laws and Customs of War on Land. Poison and poisoned weapons were unconditionally outlawed: an expression of the fundamental principles that the means of injuring the enemy are not limitless and that warfare is subject to humanitarian law.

By the end of the nineteenth century, discoveries in organic chemistry pushed industrial development in Europe and the United States forward. Fear of military exploitation of the toxic properties of some of the new compounds led to the adoption of the 1899 Hague Declaration 'Concerning Asphyxiating Gases', by which the contracting parties agreed to 'abstain from the use of projectiles the sole object of which is the diffusion of asphyxiating or deleterious gases'.[3] This reference to projectiles, however, enabled Germany to claim its first large-scale cylinder poison gas attack in April 1915 did not violate the rules of war, as no shells were involved.

In hindsight, the 1899 Hague Declaration (Article IV(2)) raised some fundamental questions regarding the impact of emerging technology on warfare and the precise meaning of the principle of humanity in war. Technology was perceived as 'value neutral' and no compelling need was felt to restrain it.[4] Moral judgement was reserved for its application in war. Consequently, the agreements of the time placed constraints on the use of certain types of weapon and not on the weapons themselves. Humanity in war also assumed a double meaning. Regulating certain modes of warfare or banning weapons that caused superfluous injuries could ameliorate the unnecessary suffering of the individual or non-combatant. Nevertheless, modern technology also offered the possibility of war so violent and destructive that fighting could only be of short duration, thereby causing fewer casualties. Humanity in war was thus transmuted into the statistic of dead, wounded and recoveries from injuries. The CWC ended most debates by de-legitimising the entire class of weapons. However, calls for the de-legitimisation of non-lethal technologies, which included incapacitating chemicals, demonstrate that the discussion shifted again.

After the First World War, the major Allied powers attempted to translate the widespread revulsion against chemical warfare into an international prohibition on the use of such weapons. The subsequent discussions led to the adoption of the 1925 Geneva Protocol for the Prohibition of the Use in War of Asphyxiating Poisons or other Gases. Until the entry into force of the CWC, the Geneva Protocol remained the sole document constraining the employment of toxic chemicals in war. However, the scope of the prohibition was limited solely to the use of chemical weapons and not to their development, production and stockpiling. Additionally, several major contracting powers attached reservations declaring that the protocol bound them only to other states that had also signed, ratified or acceded to it. They also stated that the protocol would cease to be binding on them if another contracting power or its Allies first attacked them with chemical weapons. Thus reduced to a no-first-use statement, the agreement did not remove the justification for chemical armament and preparations for chemical warfare. Furthermore, the Geneva Protocol did not contain any verification mechanisms if use was alleged, nor did it provide for sanctions in the case of a proven violation. The agreement, nevertheless, acquired great moral authority and constrained preparation for, and resort to, chemical warfare.

In the first half of the 1930s, negotiations were conducted in the League of Nations to reduce the level of armaments generally. Nevertheless, several proposals contained clauses directly pertaining to chemical weapons, to prohibit their development and production in peacetime and to destroy existing stockpiles. A special committee was set up to deal with issues such as the definition of chemical weapons, the verification of treaty compliance and the imposition of sanctions in the event of violations. In March 1933, Britain submitted a far-reaching draft treaty containing a definition of chemical weapons that included lachrymatory and incendiary agents. The new agreement would also have prohibited the use of chemical weapons against non-parties to the treaty. The right of retaliation was maintained. However, the Disarmament Conference ceased its activities in January 1936 as a result of the worsening international climate in Europe and Asia. Italy resorted to chemical warfare in Abyssinia and the international community failed to take Coherent action. Instead, military thinkers began to theorise about the awesome potential of fleets of aircraft armed with chemical bombs against enemy cities and, indeed, some European powers instituted extensive civil defence programmes. The 1930s ended with a fear of the massive employment of chemical weapons in the next war.

However, during the Second World War, apart from Japanese operations in China, chemical weapons were not used and, after the defeat of the Axis powers, the advent of the atomic bomb overshadowed chemical warfare-related issues. Chemical weapons essentially disappeared from the disarmament scene until the late 1960s when events in the Vietnam War prompted the United Nations to prioritise chemical disarmament. However, it should be noted that the United Nations is, of course, the servant of its member states, not master of them, and as such is really in no position to prioritise anything. Indeed, the reality was that in 1968 was that the Eighteen Nation Disarmament Conference decided to include chemical and biological warfare on its agenda.[5] The discussions in the 1930s had ended in failure, but the British draft had sown the seeds for the chemical weapon disarmament treaties of the latter half of the twentieth century. In particular, the British proposal heralded the shift from constraining the use of chemical weapons in war to the total abolition of a particular class of arms in peacetime.

In 1968, talks had opened on chemical and biological disarmament, but it was not until 1972, when the Biological Warfare Convention (BWC) had been signed, that negotiations on the CWC began, within the framework of the United Nations Committee on Disarmament.[6] However, an agreement on chemical weapons almost immediately proved difficult to achieve and during the late 1970s the marked deterioration in East–West relations added to the complexities of banning a proven weapon. The positions on politically sensitive issues such as the nature and extent of verification measures remained far apart. A series of bilateral negotiations between the United States and the Soviet Union in the late 1970s and early 1980s also failed to achieve a breakthrough. The slow progress of the negotiations led to the development of the idea of a chemical weapons free zone in Europe. This proposal was immediately rejected by the NATO because it would have undermined its retaliatory option. Indeed, had a chemical weapons free zone been created in Europe, NATO would have been forced to transport chemicals across the Atlantic Ocean in the event of the initiation of chemical warfare by the Warsaw Pact. In any event, the idea died silently when East–West relations improved greatly in the latter half of the 1980s and a global ban on chemical weapons became a distinct possibility. Certainly, the extensive use of chemical warfare in the Iran–Iraq War added urgency to the talks in Geneva.

In 1984 an important milestone was reached when the negotiators agreed on a basic structure of a preliminary draft treaty, based on a proposal submitted by the United States. A second series of bilateral talks between the United States and the Soviet Union between 1986 and 1991 gave impetus to the multilateral process. In particular, both countries began exchanging detailed information on their respective chemical weapons stockpiles and committed themselves to verified destruction. The United States also agreed to end the controversial programme for the production of binary chemical munitions which it had begun in 1987. The experience of the threat of chemical warfare in the Gulf War in 1991 enabled the negotiators at the Conference on Disarmament to reach a final agreement in September 1992.

On 13 January 1993 the Convention on the Prohibition of the Development, Production, Stockpiling and Use of Chemical Weapons and on their Destruction was opened for signature in Paris.[7] On 29 April 1997, almost 100 years after the first agreement restricting the use of projectiles with harmful gases was signed, the majority of the world's nations joined to activate an arms control and non-proliferation accord that was aimed to gradually compel the elimination of one of the most abhorred classes of weapons of all time. Previously, the international community had fallen short of the mark in efforts to try to abolish poison gas, despite the shame following its widespread use in the First World War. The new Chemical Weapons Convention extended the no-use prohibitions of the Geneva Protocol to ban the development, acquisition, production, transfer and stockpiling of chemical weapons as well. The overall purpose of the CWC was to prevent the possibility of the use of chemical weapons and to ensure the destruction of existing chemical weapons production facilities and arsenals over a 10-year period. In contrast to the Geneva Protocol it did not allow any reservations. States could never, under any circumstances, engage in military preparations for chemical warfare and therefore they gave up the option of in-kind deterrence or retaliation. The CWC required nations to publicly declare activities that were previously considered state secrets and private business information. The treaty also authorised routine inspections to monitor compliance with its prohibitions, a revolutionary concept in treaties.[8] Additionally, the treaty prohibited the use of riot control or anti-plant agents as methods of warfare.

The CWC consisted of 24 main articles and 3 annexes. Articles I and II outlined the basic provisions of the treaty and included definitions of the various terms used throughout. Article III obliged state parties to submit declarations of their past programmes, including information on current holdings of chemical weapons and production facilities. Article IV detailed chemical weapons and Article V dealt with production facilities. Article VI required state parties to allow a degree of verification of chemical industry facilities working with certain 'dual-use' chemicals. Article VII contained rules to facilitate the implementation of the CWC by each state party. Article VIII established the Organisation for the Prohibition of Chemical Weapons (OPCW) and defined the powers and functions of its three constituent organs – the Conference of State Parties, the Executive Council and the Technical Secretariat. Article IX detailed procedures through which state parties could resolve any questions related to non-compliance. Article X gave state parties the right to develop protective programmes against the use of chemical weapons and outlined assistance which could be provided by the OPCW, in the event of an attack by chemical weapons. Article XI stated that the CWC should not inhibit the economic and technological development of state parties or hamper free trade in chemicals and related technology and information. Article XII included measures to redress a situation of non-compliance, including sanctions. The remaining 12 articles were shorter and dealt with legal issues such as the CWC's relationship with other international agreements, settlement of disputes, amendments, duration and withdrawal.

The three annexes – on chemicals, on implementation and verification, and on the protection of confidential information – were an integral part of the CWC. The annex on chemicals listed, in three schedules, 43 chemicals and families of chemicals that were selected for the application of special verification procedures. Schedule 1 contained toxic chemicals that had been manufactured as chemical weapons or their key precursors (in other words, chemicals that could be used as a stage in their synthesis) and for which there were no peaceful uses, for example, the nerve agents and mustard gases. The list also included two toxins, saxitoxin and ricin. Schedule 2 contained chemicals which could be used as precursors but which also had relatively limited use for non-prohibited purposes; for example, thiodiglycol, which not only is a precursor for mustard gas but is also widely used as a solvent in printing inks. Schedule 3 contained chemicals, such as phosgene and hydrogen cyanide, which had been used as weapons, but along with precursors were used in large quantities for civil chemical industry purposes. The annex on implementation and verification procedures provided great detail on the conduct of the CWC's verification provisions, from declarations and inspections to challenge inspections and investigations of alleged use.

The foundation of the CWC's inspection activities was based around the declaration by member states of their chemical weapons capabilities and activities. Nations with chemical warfare programmes were required to declare their production, storage and destruction facilities, which would then receive top monitoring priority. Nevertheless, the CWC did allow states to maintain research programmes to ensure the integrity of defensive equipment such as gas masks and gas detectors, but these activities were also to be closely monitored since they involved work with the chemical agents listed on Schedule 1.[9] Otherwise, all other warfare agents, mustard gas, Lewisite, soman, sarin, tabun, VX and the capability to produce them were to be eliminated under the watchful eyes of international inspectors (Table 8.1).[10] The convention thus defined chemical weapons as any toxic chemical, or its precursors, intended for purposes other than those not prohibited under this convention for

Table 8.1 Timeframe for chemical weapons destruction under CWC

Percentage of stockpile destroyed	Year after entry into force
Planning/testing	1–2
1	3
20	5
45	7
100	10

Source: CWC, Verification Annex, Part IV(a), paragraph 17(a).

research, as well as munitions, devices or equipment specifically designed to be used with them.

The CWC's other declaration requirements pertained to the numerous chemicals used in ordinary commercial products – fertilisers, pesticides, flame retardants and pharmaceuticals – that also could be diverted to manufacturing poison gas.[11] Chemicals that industry used widely for commercial products were listed in Schedule 3,[12] and the chemical industry was required to declare its activities with controlled chemicals above certain threshold quantities. However, the CWC was not restricted to compounds that were explicitly listed in the schedules. The discovery of a new potential chemical warfare agent would not, therefore, undermine the CWC because such an agent would be automatically banned if it had no justifiable non-military purpose.

The CWC's on-site inspections were designed to acquire factual evidence to confirm whether or not the CWC's members were engaged in legitimate commercial and defence activities and not weapons programmes. If an inspection team found factual evidence inconsistent with declarations, or host officials could not sufficiently clarify other ambiguities uncovered during the course of an inspection, the inspectors were empowered to use more intrusive methods to determine whether chemical agents were being produced on-site and whether chemicals were being diverted for chemical agent production.[13] Given the huge number of commercial chemical plants worldwide, the CWC inspectors would have been unable to check frequently on every facility capable of making chemical weapons and so the CWC strategy was to place inspectors routinely at the facilities with the highest proliferation risk, and randomly at other industry sites.

When the CWC entered into force, more than 20 countries were thought to possess offensive chemical weapons capabilities. The countries thought to be of proliferation concern in the late 1990s were China,

Egypt, India, Iran, Iraq, Israel, Libya, North Korea, Pakistan, Russia, South Korea, Syria, Taiwan and Vietnam.[14] Table 8.2 illustrates the proliferation threat in 2004, in terms of both the chemical agents and the delivery systems these countries are believed to possess.

Table 8.2 Chemical proliferation threat 2004

Country	Chemical weapons capabilities
China	Has advanced chemical weapons programme. Researching more advanced agents.
	Delivery systems include artillery, rockets, mortars, landmines, aerial bombs, sprayers, and short- and medium-range ballistic missiles.
Egypt	Probably maintains a chemical weapons stockpile.
India	Declared in June 1997 that it possessed a chemical weapons stockpile. Has begun to destroy its stockpile under the CWC. Its industry retains the ability to produce agent precursors – chemicals that can be used in chemical weapons production.
	Delivery systems include short-range anti-ship cruise missiles, air-launched tactical missiles, fighter aircraft, artillery and rockets.
Iran	Has a stockpile of chemical weapons. Previously known to have produced and stockpiled blister, blood and choking agents, and probably nerve agents. Seeking aid from Chinese and Russian entities to develop more advanced self-sufficient infrastructure.
	Delivery systems include artillery shells, mortars, rockets and aerial bombs.
	Used chemical weapons during Iran–Iraq War.
Iraq	Had extensive programme before the Gulf War under which it produced and stockpiled mustard gas, sarin, tabun and VX.
	Delivered chemical agents against Iranian forces during Iran–Iraq War using aerial bombs, artillery, rocket launchers, tactical rockets and helicopter-mounted sprayers. Also used chemical weapons against own Kurdish population in 1988.

Table 8.2 (Continued)

Country	Chemical weapons capabilities
	Programme largely dismantled by United Nations weapons inspectors after Gulf War but Iraq retained some chemical weapons and began reconstituting its chemical infrastructure after inspectors left the country in 1998. Could resume agent production but would need foreign assistance to completely restore its production capabilities to pre-Gulf War levels.
	Reports suggest that in mid-2001 Iraq had stocks to produce significant quantities of mustard gas within weeks and nerve agents in months. The fact that no chemical weapons or agents were discovered following the war in 2003 suggests the Joint Intelligence Committee (JIC) assessment for the British and American governments was incorrect and, indeed, United Nations inspections and destruction policies were successful following the Gulf War.
	Some additional potential delivery systems (as well as those used in Iran–Iraq War) include short-range ballistic missiles, fighter aircraft and un-manned aerial vehicles.
Israel	Probably has a chemical weapons programme.
Libya	Produced mustard gas and nerve agents before 1990. Working to re-establish chemical weapons capabilities that had been limited by United Nations sanctions from 1992 to 1999. Highly dependent on foreign supplies.
	Attempted to use chemical weapons against Chad troops in 1987.
	Potential delivery systems include short-range anti-ship cruise missiles, air-launched tactical missiles, fighter aircraft, bombers, artillery, helicopters and rockets.
North Korea	Believed to possess sizeable stockpile of chemical weapons including nerve, blister, choking and blood agents.
	Delivery methods could include missiles, artillery and aerial bombs.

Pakistan	Has imported chemicals that it could use to make chemical weapons agent.
	Delivery methods could include missiles, artillery and aerial bombs.
Russia	Possesses about 40,000 metric tons of chemical agent including sarin, soman, mustard gas, VX, Lewisite and phosgene.
	The United States believes Russia has not declared some of its chemical agents and weapons, and notified Moscow (April 2002) it could not certify that Russia was complying with the CWC.
	Has started destroying its chemical weapons under the CWC but is not expected to complete the destruction until at least 2012.
	Reports suggest that Russia has worked on a new generation of chemical agents called 'novichoks', which are allegedly designed to circumvent the CWC and evade Western methods to detect and protect against chemical weapons.
	Potential delivery systems include artillery, bombs, spray tanks and short-range ballistic missiles.
South Korea	Possesses a chemical weapons stockpile and is destroying it under the CWC.
Syria	Possesses sarin which it can deliver by aircraft or ballistic missile, and is working to develop VX. Key elements of its programme rely on foreign sources.
United States of America	Possesses about 31,000 metric tons of chemical agent including sarin, VX and blister agents. Is currently destroying its stockpiles under the CWC.
Federal Republic of Yugoslavia	Possesses weaponised CS. Suspected of having unweaponised mustard gas and sarin, and possibly weaponised BZ.

Sources: Arms Control Association; Monteray Institute of International Studies; 'Chemical and Biological Weapons: Possession and Programmes Past and Present', http://cns.miis.edu/research/cbw/possess.htm; Russian Government Resolution No. 510 (5 July 2001); *Iraq's Weapons of Mass Destruction: The Assessment of the British Government*, www. pm.gov.uk.

Chemical weapon proliferation is usually described as a lateral spread of the precursor chemicals, dual-use high technology and expertise from developed to developing countries. The issue came to the fore in 1984 when it became clear that Iraq was using chemical agents in its war with Iran. Moreover, it was soon realised that companies from the developed world were knowingly, or unknowingly, involved in Iraq's chemical weapons programme. To prevent this happening again the governments of Western countries set up national export control policies and began to co-ordinate their efforts in the Australia Group from 1985.

In January 1989, as the world's leaders met in Paris to restore the authority of the Geneva Protocol following the Iran–Iraq War, global attention focused on Libya's large chemical warfare factory at Rabta. It became clear that west European companies, with the assistance of some firms in eastern Europe and Asia, were deeply involved in the construction of the plant despite the existence of export controls. These events, together with the chemical warfare threat during the Gulf War, caused governments of the industrial world to advocate a more permanent non-proliferation regime to supplement the CWC.

It is difficult to assess the global chemical weapons proliferation threat. New information about chemical weapon armament programmes in some countries, nevertheless, leads to the conclusion that approximately 13 per cent of all nations are believed to have engaged in some form of chemical weapon armament. During the First World War reliable evidence indicates that 17 per cent of all nations possessed chemical weapons; the figure for the Second World War was 19 per cent.[15] Comparisons may be misleading, however, because publicly available reports do not define chemical weapons capability. Despite the apparent increase in the number of proliferators, the mix of chemical weapons possessors may vary at different times. Assessments are further complicated by the indigenous acquisition of knowledge, expertise and technologies by developing countries as part of their legitimate industrialisation programmes.

Pakistan, India and China were among several states that adamantly disclaimed a chemical weapons capability before joining the CWC, but India and China have now opened their chemical weapons programmes to the scrutiny of inspectors following the CWC's activation. The CWC differentiated between two different types of offensive capabilities: poison gas arsenals and production facilities. India and South Korea declared both chemical weapons plants and stockpiles. Two other countries also made the same declarations, the United States and Russia, which were known chemical weapons possessors before the CWC entered into force. Seven states, China, France, India, Japan,[16] Russia, the United Kingdom and the United States, declared former chemical weapons production facilities. At first, this inventory of states declaring chemical weapons capabilities might appear modest, but several states are believed to have quietly destroyed their chemical arsenals and production facilities in preparation for the CWC's entry into force. France and China, for example, are thought to have taken this course of action. Other countries have recently declared chemical weapons and/or chemical weapon production facilities, including Bosnia, Serbia, Albania and Libya.[17]

The United States received the most CWC inspections following their declaration of 10 storage facilities, seven former production facilities and six destruction sites. Incineration was the US Army's favoured destruction technology, but some parties were constantly considering alternative technologies to incineration because of opposition. Additionally, the US Army was required by law to consider alternative destruction policies from private industry. Russia declared 24 former chemical weapons facilities, 7 storage sites and 40,000 metric tons of chemical agents.[18] Initial inspections were completed at all the Russian locations, but the financial crisis in Russia following the collapse of the Eastern bloc in 1990/1991 slowed down efforts to get a destruction programme underway. However, the destruction efforts were not only hindered by the financial crisis but also by a failure by the Russian Duma to allocate funds, earmarked for chemical weapons destruction, provided by Germany, the Netherlands, Sweden and the United States, and by local and federal opposition to the draft destruction plan. Hearings on chemical weapons destruction held by the Duma Committee on the Environment also demonstrated that a number of fundamental aspects of destruction, including the choice of destruction technologies, were either unfamiliar or objectionable to a significant number of those who spoke during the hearings. Nerve agents comprised 81 per cent of Russia's stockpile that Moscow finally proposed to destroy in a two-step process: neutralisation followed by bituminisation.[19] Neutralisation is a process whereby a chemical is inserted into the agent to dilute its potency significantly. Russia proposed to use monoethanolamine to neutralise sarin and soman, and potassium isobutylate to neutralise VX. Bituminisation further dilates the toxicity of the resulting waste product by mixing it with asphalt and calcium oxide hydrate. The two-step process produces insoluble salts which, in Russia's case, will probably be stored indefinitely.

The difficulty of getting destruction programmes underway aside, another obvious shortcoming in the CWC's implementation was that the treaty was not universal. By 1999, 18 months after it came into force, only 150 countries had joined the CWC and more than 50 countries signed it but did not ratify it. More worrying, by virtue of their absence, were the suspected proliferators including Egypt, Iraq, Israel, Libya, North Korea, Syria and Vietnam. Vietnam joined in 1998 but the remainder are still absentees.[20] The treaty made it clear that absentee states would be cut off from supplies of chemical weapons precursors because all CWC members were required to enact export controls precluding such trade with them.[21] Besides damaging economies and standards of living in these countries, the penalties, it was envisaged, would make it more difficult to sustain a chemical weapons programme.

However, the CWC will not constitute more than a mere piece of paper unless members today regard it as a vital mechanism to help ensure national security and international peace. If the member states consider the act of joining the treaty and the process of meeting its declaration and inspection requirements as a measure of the treaty's success, then they will have taken for granted the treaty's tools to determine and enforce compliance. Neglecting to police compliance with the CWC will extend the treaty's behavioural norm against chemical weapons possession, just as the Geneva Protocol's ban against the use of poison gas was diminished when the international community took no punitive measures against Iraq, a flagrant transgressor of the Geneva Protocol in the mid-1980s. Indeed, initially governments weakened the Geneva Protocol by placing significant reservations on it when they ratified it, in effect downgrading it to a no-first-use treaty. The behavioural norm against chemical weapons was further subdued when Iraq used chemical weapons with impunity during its war with Iran and the international community did not enforce the protocol. Indeed, at a special conference convened in Paris in 1989 to consider what to do about this blatant violation, 149 countries in attendance condemned the use of chemical weapons, but did not even name Iraq as a violator of the Geneva Protocol! However, of course, the purpose of this conference was not to vilify Iraq but to reaffirm the protocol's prohibitions, especially by Arab League counties, which was indeed what happened.

Until member states make the effort to review declarations and demand access to the final reports from inspections, they will be in no position to ascertain whether other states have accurately declared their military and civilian chemical capabilities. Nonetheless, the declarations are reviewed avidly by those member states that take chemical weapons seriously. The CWC denies access to the raw inspection reports, but not to compilations of data from them, which, again, are read avidly. However, given the fact that members suppress their own inspection reports, perhaps it is not surprising that no state has requested a challenge inspection of another. The treaty's challenge inspections were designed to detect and deter non-compliance. No other treaty contains such a strong inspection tool, including the obligation for CWC members to co-operate with a challenge inspection but, of course, it is not known how such an inspection would work in practice. Some policymakers expect a challenge inspection team to return with irrefutable evidence of non-compliance, others fear that challenge inspections will be little more than overt spying expeditions, and still others think the whole concept will flop. The truth probably lies somewhere in between. While there is no guarantee that an inspection team would uncover absolute evidence of cheating, the results of an inspection would, undoubtedly, shed more light on what has happened at a suspect site.[22] One thing, however, is certain: detection and deterrence are unlikely to occur if the inspection tool remains dormant. Yet, no government to date has activated the CWC's Article IX investigative procedures and such behaviour helps subvert the challenge inspection process to a concept that members think of in merely hypothetical terms.

All members share the responsibility for enforcing the treaty's bans and, yet, in the United States policymakers and citizens believe that their country has by several measures, not least by virtue of an unparalleled military strength and the world's largest economy, earned a singular place in the international community.[23] However, historically, the United States has set a poor example under the CWC. It has infringed the treaty, failed to pay its bills, failed to deliver promised equipment, undertaken corrupt conduct during inspections and drafted legislation containing treaty-weakening exemptions. America's lifeless performance in the CWC arena is all the more inconsistent given the energy with which the United States supported the UNSCOM's efforts to oversee the elimination of Iraq's WMD capabilities and, more recently, they have sought to prevent terrorists from acquiring chemical weapons. The United States emphasised the importance of free and unfettered access to UNSCOM's ability to track down the remnants of Iraq's WMD programmes, wherever they may have been hidden. Free and unfettered access is to UNSCOM what challenge inspections are to the CWC. Prior to the war on Iraq in 2003, UNSCOM evidence of threatening chemical weapons activity in Iraq came from missile fragment samples, one of which proved Iraq weaponised the nerve agent VX. The subsequent United Nations Report stated: 'Significant amounts of VX disulphide ... and stabiliser were found in the samples.'[24] Partly, on such evidence, a war was launched. Furthermore, the United States launched a cruise missile attack against a pharmaceutical plant in the Sudan based on a soil sample analysis which the US government held as irrefutable proof that an international terrorist organisation had manufactured a VX precursor at this plant. Of the soil sample, and the US case against terrorist activities sponsored by Osama Bin Laden at the al Shifa Pharmaceutical Plant outside Khartoum, US National Security Advisor, Samuel R. Berger, stated: 'There is no question in our minds that that facility, that factory, was used to produce a chemical that is used in the manufacture of VX gas and has no other commercial distribution as far as we understand.'[25] The chemical in question was later identified as ethyl methyorthophosphric thionate (EMPTA), used to treat seeds and turf grasses.

When all is said and done the United States took a very nonchalant approach to the implementation of the CWC and, yet, throughout the treaty's negotiation the United States was one of the strongest champions of a chemical weapons ban, persuading other countries to conclude an agreement. Washington extolled the CWC as the centrepiece of international efforts to reduce the chemical weapons threat, but since signing the treaty in 1993 the United States has left the CWC virtually untended. Other nations have closely observed America's treatment of the CWC. With the world's second largest chemical weapons stockpile and the largest commercial chemical industry, the United States is perceived by the international community to have special responsibilities regarding the CWC's implementation. Supporters of the CWC hope that the United States will regain its focus as a steadfast advocate of the treaty; foes of the CWC stand ready to contribute to its downfall should the United States not redeem itself.

Chemical weapon disarmament has progressed far since the first attempts were made a 100 years previously to outlaw the use of chemical weapons in war. The CWC still holds the best promise for reducing the threat of chemical warfare by building an environment of confidence and security. As well as instruments of verification and inspection, the CWC also possesses resort to aid and assistance in the area of chemical warfare defences in case of attack. In the final analysis, however, the overriding aim of the CWC continues to be the effective ban of all chemical weapons, complemented by the desire to promote the peaceful use of chemicals in industry. Whether such incentives will balance the pressure to acquire such weapons remains to be seen.

Notes

Preface

1. www.globalsecurity.org/wmd/library/1984/ARW.htm, p. 3.
2. Fritz Haber, Nobel Acceptance Speech, www.nobel/prizes/1919.htm.
3. Victor Lefebure, *The Riddle of the Rhine*, London (1919), p. 2.
4. Stockholm International Peace Research Institute (SIPRI), *The Problems of Chemical and Biological Warfare*, Vol. 1, Stockholm (1971), p. 50.
5. Matthew Meselson, 'The Yemen', in Stephen Rose (ed.), *Chemical and Biological Warfare*, George G. Harrap & Co. Ltd, London (1968), p. 101.
6. SIPRI, op. cit., Vol. 1, p. 44.
7. J. Perry Robinson *et al.*, *World Health Organisation Guidance*, Geneva (2004).
8. J.J. Pershing, *Final Report of General John Pershing*, US Government Printing Office: Washington DC (1920), p. 77.

1 Historical precedents?

1. http://www.un.org.
2. SIPRI, Vol. 1, p. 25.
3. Edgewood Arsenal, 'Status Summary of the Relative Values of AC, CK and CG as Bomb Fillings', Project co-ordination Staff Report No. 1 (10 July 1944).
4. J.R. Wood, 'Chemical Warfare – A Chemical and Toxicological Review', *American Journal of Public Health*, 34 (1946), pp. 455–460.
5. Anon., 'BC Stridsmedel', *Forsvarets Forkninganstalt*, No. 2 (1964).
6. For a fuller discussion of Britain's discovery of VX see, Caitriona McLeish, 'The Governance of Dual-Use Technologies in Chemical Warfare', MSc Dissertation, SPRU, University of Sussex (1997), especially pp. 50–58.
7. The 12th Earl of Dundonald, *My Army Life*, London: Arnold (1926), p. 330.
8. See *Studies*, G.B. Grundy (1911, 1948); J.H. Finlay (1942) and A.W. Gomme (1945).
9. Von Senfftenberg, *Von Allerei Kriegsgewehr und Geschütz* (mid-Sixteenth Century), source located in British Library, London.
10. A chemical weapon is the agent and the projectile; the chemical itself is referred to as a chemical warfare agent.
11. SIPRI, Vol. 11, p. 126.
12. Dundonald, *Army*, op. cit., p. 147.
13. *Ibid.*, p. 153.
14. *Ibid.*, p. 54.
15. http://www.tripod.com/Napchemxxx.htm.
16. *Ibid.*
17. Dundonald, *Army*, op. cit., p. 279.
18. Radhika Lexman, 'Understanding the Threat', *Kuwait Samacher* (2003).
19. http://www.stpetersburg/text/htm.

20. Alden H. Waitt, *Gas Warfare. The Chemical Weapon. Its Use and Protection Against It*, New York: Duell, Sloane & Pearce (1942), pp. 7–11.
21. The Hague Declaration, Clause IV/2.
22. Waitt, *Gas*, op. cit., pp. 12–13.

2   The First World War

1. Wilfred Owen, *Dulce et Decorum Est* (It is a wonderful and great honour to die for your country). Reproduced in full in, David Roberts, *Minds at War*, Saxon Books (1996).
2. From the diary of Lance-Corporal A. Barclay, 1/1st Ayrshire Yeomanry, kindly loaned to the author by Jonathan Menhinick.
3. www.imperialservices.org.co.uk/military_tactics/html.
4. *Nature* (12 January 1922).
5. In First World War terms, chemical warfare included not only gas, but liquid flammable material, thermite and smoke. Chapter 2 only deals with what participants referred to as chemicals, gases or war gases. These included real gases such as phosgene and chlorine and also weapons that, while referred to as gases, were in fact vaporised liquids – mustard gas for example – or finely ground solids.
6. Geoffrey Regan, 'Blue on Blue', *The Historian*, www.thehistorian.co.uk.
7. SIPRI, Vol. 1, p. 125.
8. *Ibid.*, p. 126.
9. For a detailed discussion of this shell type see, Anon., *Army and Navy Journal* (8 May 1915), p. 1141.
10. T-Stoff was a mixture of brominated aromatics including xylyl bromide, xylylene bromide and benzyl bromide. All these compounds are extreme irritants capable of severely limiting the effectiveness of unprotected troops.
11. Eric von Ludendorff, *War Memories*, Berlin (1919).
12. The interesting point about Haber's work on chemical warfare is that he did it on his own initiative. In fact, he approached the German military at the end of 1914 to sell them on poison gas, but the military had no great respect for scientists and poison gas seemed 'unsporting anyway'.
13. *Chemical Trade Journal*, 65, No. 1677 (1919), p. 37.
14. H. Schwarte, *Die Technik Im Weltkrieg*, Berlin: Mittler (1920), p. 74.
15. SIPRI, Vol. 11, Chapter 1.
16. Rudolf Hanslian, *Der deutsche Gasangriff bei Ypern am 22 April 1915*, Berlin: Verlag Gasschultz und Luftschutz (1934), pp. 10–12.
17. PRO, WO 158/815, Chlorine Gas (June 1915).
18. Elmer Cotton in Malcolm Brown, *Tommy Goes to War*, J.M. Dent & Sons (1978), p. 170.
19. PRO, 30/57/80, SUPP 10/92, Reports on the use of Poison Gas (1918).
20. Hanslian, op. cit., p. 64.
21. www.users.globalnet.co.uk/dccfarr/gas.htm.
22. Hanslian, op. cit., p. 186.
23. Sir J.E. Edmonds, *Military Operations: France and Belgium*, London, Vol. iii (1927), pp. 163–164.

24. On 17 December 1932 Auguste Jaeger was tried at Liepzig on charges of desertion and betraying German plans to the enemy. He was sentenced to ten years imprisonment.
25. Militärarchiv der Deutschen Demokratischen Republik (MA), W31.60/2021 (June 1915).
26. Victor Lefebure, *The Riddle of the Rhine*, op. cit.
27. Hanslian, *Der Chemische Krieg*, op. cit., 2nd edn (1927).
28. Rudolf Hanslian, *Gasangriffe*, op. cit.
29. W.G. McPherson *et al.* *[Official] History of the Great War: Medical Services*, Vol. 2, 'Diseases of the War', London (1923).
30. Hanslian, *Der Deutsche Gasangriff*, op. cit.
31. C.H. Foulkes, *Gas! The Story of the Special Brigade*, London (1936).
32. Hanslian, *Der Deutsche Gasangriff*, op. cit.
33. *New York Tribune* (26 April 1915).
34. *New York Tribune* (27 April 1915).
35. Hansard (Lords) 208 (11 June 1915).
36. MA, W31.60/2021 (June 1915); PRO, WO 32/5169, Operations: General Code 46 (A) (June 1915).
37. PRO, WO 32/5169, Operations: General Code 46 (B) (June 1915).
38. PRO, WO 32/5173, Engineers Code 14 (E): Formation of special poison gas companies, Royal Engineers (1915). The special gas companies were later to evolve into the Royal Engineers Special Brigade.
39. PRO, WO 142/265, The Royal Engineers Experimental Station, Porton (1919).
40. LHMA, FOULKES 6/37, Establishment of Central Laboratory, Helfaut (1915).
41. C.B. Carter, *Porton Down: 75 Years of Chemical and Biological Research*, London: HMSO (1992), p. 10.
42. PRO, WO 32/595, Casualties caused by poison gas in British Forces 1915–1918 (1918–1919).
43. German Sixth Army War Diary, Extracts in Rudolf Hanslian, *Der Chemische Krieg*, Berlin: E.S. Mittler und Sohn (1927), pp. 86–98.
44. Regan, op. cit., p. 5.
45. When the Ministry of Munitions was established in Britain in June 1915 the production of high-explosive shells were 92 per cent in arrears. Throughout the First World War this arrears was never rectified.
46. LHMA, FOULKES 6/6, Papers relating to the Battle of Loos on 25 September 1915 and 'Sir John French's Dispatch' (15 October 1915).
47. The development of mustard gas resulted from pure scientific investigation from as early as 1860. Victor Meyer, a German chemist, described the substance in 1884, indicating its skin blistering effects. There is some evidence of further investigation in German laboratories in 1913, and whatever motive for this work, we know that mustard gas must have received the early attention of the German War Office, for it was approved and was in production early in 1917.
48. PRO, WO 32/5176, Report on Mustard Gas attack by Germany (1917); PRO, WO 106/148, Reports on Mustard Gas (April 1917 to September 1919).
49. See Table 2.2.

50. PRO, WO 32/5177, Laws of War: Code 55 (E), Threat of use of new poison gas shells (1918–1919); PRO, MUN 4/1684, Gun ammunition filling (2 May 1918).
51. T. Watkins, J.C. Cackett and R.G. Hall, *Chemicals, Pyrotechnics and the Fireworks Industry*, Oxford: Permagon Press (1968), p. 9.
52. *Ibid.*, p. 10.
53. Ulrich Miller-Kiel, *Die Chemischer Waffe im Weltkrieg und Jetzt*, Berlin: Verlag Chemie (1932), p. 49.
54. PRO, WO 142/71, WO 142/72, Chemical War Committee Minutes 77–112 (23 October 1917 to 28 March 1919); WO 188/114, Chemical Warfare – Germany (1918–1922).
55. PRO, MUN 4/2735, Chemicals and their use by German and British armies (10 August 1917).
56. Hanslian, *Chemische*, op. cit., pp. 48–49. It is interesting to note that as early as 1887 Professor Bayer, a renowned organic chemist, in his lectures at the University of Munich, included a reference to the military value of lachrymators.
57. *Ibid.*, p. 50.
58. *Ibid.*, p. 75.
59. Pierre Berton, *Vimy*, Toronto: McClelland & Stewart (1986), p. 115.
60. LHMA, FOULKES 6/5, Papers relating to the Livens Projector (1915–1921).
61. Hanslian, *Chemische*, op. cit., p. 107.
62. Major Charles E. Heller, USAR, *Chemical Warfare in World War I: The American Experience, 1917–1918*, Combined Arms Research Library, Command and General Staff College, Fort Leavenworth, Kansas, Leavenworth Papers (hereafter cited as Leavenworth Papers), No. 10, p. 17.
63. *Ibid.*, p. 18.
64. *Ibid.*, p. 17.
65. LHMA, FOULKES 6/45, Official evidence complied from Germany of the results of British gas attacks (1918–1920).
66. Leavenworth Papers, No. 10, p. 20.
67. PRO, WO 142/71–72, Chemical Warfare Committee Minutes (27 October 1917 to 28 March 1919)
68. Hanslian, *Chemische*, op. cit., p. 45.
69. A.M. Prentiss, *Chemicals in War: A Treatise on Chemical Warfare*, New York: McGraw-Hill (1937), p. 58.
70. MA, W31.60/20.
71. Haber, *Poisonous Cloud*, op. cit., p. 165.
72. For more information on German casualty figures see, W.G. Macpherson *et al.* *[Official] History of The Great War: Medical Services: Diseases of the War*, London (1923), Vol. 2, especially Chapter 9.
73. PRO, WO 33/1014, 2nd Annual Report of the Chemical Defence Research Establishment, Porton Down (1922).
74. PRO, WO 33/1049, 4th Annual Report of the Chemical Defence Research Establishment, Porton Down (1924).
75. J.D.S. Haldane, *Callinicus: A Defence of Chemical Warfare*, London (1925), p. 74; Haldane also describes in this book how 'someone placed a drop of mustard gas on the chair of the Director of the British Chemical Warfare Department. He ate his meals off the mantelpiece for a month!'

76. PRO, WO 142/96–98, Mustard Gas files, DES/1/–80, DES/2/81–189; DES/4/281–390; DES/5/391–489.
77. SIPRI, *The Problem of Chemical*, op. cit., Vol. 1, p. 130.
78. Op. cit., Prentiss, *Chemicals*, pp. 661–666.
79. American humorist Will Rogers cited in H.E. Galarraga, 'The Evolution of the Protective Mask for Military Purposes: Inception to World War I', *Greenwich Journal of Science and Technology*, 2(1) (2001), pp. 37–48.
80. *Ibid.*, p. 39.
81. *Ibid.*, p. 42.
82. Leavenworth Papers No. 10.
83. Robert Graves, *Goodbye to All That*, New York: Doubleday (1957), p. 95.
84. Wilhelm Hasse-Lampe, *Handbuch für das Grubenrettungswesen (International)*, Bande 1, Lübeck: Dräger (1924), p. 80.
85. For further details of the mine galleries dug under the front-line see, A. Barrie, *War Underground*, Staplehurst: Spellmount (1962, 2000).
86. The HSS-Gerät is listed in the US Army's *German Chemical Warfare Material* (1945), filed in PRO, WO 208/3025.
87. LHMA, FOULKES 6/19, Papers relating to filter fans and blowers for dug-outs (1917).
88. LHMA, FOULKES 6/1–81, Papers relating to the introduction and use of gas warfare during the First World War (1915–1919).

3 The inter-war years, 1919–1939

1. A.A. Fries., 'Sixteen Reasons Why the Chemical Warfare Service must be a Separate Department of the Army', *Chemical Warfare* (1920), Vol. 2(1), p. 4.
2. *Ibid.*, p. 7.
3. *Ibid.*, p. 3.
4. Treaty of Versailles. Article 171: 'The use of asphyxiating, poisonous or other gases and all analogous liquids, materials or devices being prohibited, their manufacture and importation are strictly forbidden in Germany.'
5. Nobel Acceptance Speech, www.nobel.se/laureates/lecture.html.
6. Pepper spray is not, as is often thought, CS, but is based on a substance 'capsaicin' that is found in chilli peppers.
7. L.P. Brophy and G.J.B. Fisher, *The Chemical Warfare Service: From Laboratory to Field*, Washington DC: Office of the Chief of Military History (1959), pp. 2–27, 49–76.
8. Victor Lefebure, op. cit., *The Riddle*.
9. *Ibid.*, p. 2.
10. *Ibid.*
11. *Ibid.*, p. 241.
12. *Ibid.*, p. 267.
13. Churchill Papers (CHAR) 20/23/13–82, Papers relating to gas attack, 1919–1920.
14. *Ibid.*
15. *Ibid.*
16. *Ibid.*
17. CHAR 20/16/1–3, Use of gas in Iraq, 1919.

18. Wing-Commander Arthur Harris, later known as 'Bomber Harris', speaking in 1920.
19. PRO, AVIA 10/338, War Cabinet Minutes, 1914 onwards.
20. The French troops were led by Marshal Pétain, the First World War hero of Verdun later discredited by his collaboration with Hitler during the Second World War.
21. *L'Humanité* (20 December 2000).
22. Although the US Senate Foreign Relations Committee favourably reported on the protocol in 1926, there was strong lobbying against it, and the senate never voted on it. After the Second World War, President Truman withdrew it from the senate together with other inactive older treaties. Little attention was paid to the protocol for several years thereafter.
23. M. Duffield, 'Ethiopia: The Unconquered Lion of Africa', *Command Magazine* (1990), Vol. 4, p. 78.
24. Quoted in H.H. Brett, 'Chemicals and Aircraft', *Chemical Warfare Bulletin* (1937), Vol. 23, pp. 151–152
25. Piers Brandon, *The Dark Valley*, Cambridge (1938), p. 74.
26. D.K. Clark, *Effectiveness of Toxic Weapons in the Italian–Ethiopian War*, Bethesda: Operations Research Office (1959), p. 20.
27. *Ibid.*, p. 20.
28. J.C. Fuller, 'Chemicals in Ethiopia', *Chemical Warfare Bulletin* (1936), Vol. 22(3), p. 85.
29. B.H. Liddell-Hart, 'The Abyssinian War', *Ordnance Magazine* (1937), p. 81.
30. *Daily Telegraph*, 'Evelyn Waugh as controversial now – 100 years after his birth – as ever', 23 April 2003.
31. PRO, WO 33/1484, Annual Report (1937).
32. SIPRI, Vol. 11 (Stockholm), p. 146.
33. *The Times* (19 August 1936), p. 10.
34. *The Times* (8 September 1936), p. 12.
35. *The Times* (4 December 1936), p. 16.
36. SIPRI, Vol. 1, p. 194; 5250th Technical Intelligence Company, *The Use of Poison Gas by Imperial Japanese Army in China, 1937–1945*; Tokyo: TIC (1946).
37. A.A. Stepanov, *Chemical Weapons and Principles of Antichemical Defence*, Moscow (1962).
38. Office of Chief Chemical Officer, GHQ, AFPAC, General Organisation Intelligence Report on Japanese Chemical Warfare, Vol. 1, Tokyo (May 1946).
39. *Ibid.*
40. BIOS, Report 714, 'The Development of New Insecticides and Chemical Warfare Agents', (no date) p. 21.
41. BIOS, Report 542, Interrogation of German CW Personnel (1946).
42. BIOS, Final Report 714, op. cit.
43. *Ibid.*
44. BIOS, Report 714, op. cit., p. 28.
45. The new agent was named 'sarin' as an acronym of the names of the four key men involved in its production: Schrader, Ambros, Rüdriger and van der Linde.
46. SIPRI, Vol. 1, p. 73.
47. PRO, WO 33/078, Annual Report (1922); WO 33/1049, Annual Report (1924); WO/1078, Annual Report (1925); WO 33/1128, Annual Report (1926);

WO 33/1153, Annual Report (1927); WO 33/1204, Annual Report (1928); WO 33/1231, Annual Report (1929); WO 33/1272–1568, Annual Reports of Chemical Defence Research Department (1930–1939). These reports are located in the PRO (now the National Archives) in O and A papers WO33.

48. The United Kingdom was unique in providing such respirators, at no cost, to the whole of its population. The actual cost to the Exchequer during the 1940s was about 2/6 d (12.5 p) for each respirator.

49. Brophy and Fisher, *Chemical Warfare*, op. cit., p. 98.

50. *Ibid.*, p. 30.

51. *Ibid.*, p. 104.

52. Office of the Chief, Chemical Warfare Service, *New Gas Detector Kit*, Washington DC: CWS (12 September 1944), Information.

53. Foulkes, *Gas*, op. cit.

54. *Ibid.*

55. *The Times* (3 June 1915).

56. H. Jones, *The War in the Air*, Oxford, Vol. 5 (1935).

57. *The Times* (3 July 1926).

58. G.H. Quester, *Deterrence Before Hiroshima*, New York (1966).

59. F.J. Brown, *Chemical Warfare: A Study in Restraints*, Princeton (1968).

60. See, for example, H. Liepmann, *Death from the Skies: A Study of Gas and Microbiological Warfare*, London (1937); N. Angell, *The Menace to our National Defence*, London (1934).

61. P. Noel-Baker, *Disarmament*, London (1926).

62. *The Times* (14 March 1921), p. 11.

63. A.A. Fries, 'Chemical Warfare Past and Future', *Chemical Warfare*, 5(1) (1920), pp. 4–5.

64. Anon., 'Chemical Warfare Bombing Tests on Warships', *Chemical Warfare*, 6(10) (1921), p. 6.

65. W. Irwin, *The Next War: An Appeal to Common Sense*, New York (1921).

66. *The Times* (2 January 1933), p. 18.

67. SIPRI, Vol. 1, p. 145.

68. H. Swyter, *Proceedings of the Conference on Chemical and Biological Warfare* (25 July 1969), American Academy of Arts and Sciences, Salk Institute, Brookline, MA (1969).

4   The Second World War

1. Albert Speer, *Inside the Third Reich*, London: Phoenix (1995), p. 552.

2. *Baltimore Sun* (12 November 1941), p. 37.

3. Imperial Chemical Industries Head Office Archives (ICHO)/DIR/1964, KSK Production (April 1927).

4. ICHO/DIR/0019, History of Imperial Chemical Industries Dyestuffs Division, typed manuscript (1942).

5. ICHO/DIR/0365, Government Construction Contracts (31 December 1943).

6. Nuremberg Industrialists Documents (NI)-6788, Affidavit by Otto Ambros (1 May 1947).

7. M. Sartori, 'New Developments in the Chemistry of War Gases', *Chemical Review*, 48 (1951), pp. 225–257.

8. Stuart Laing (ed.), *Hitler Diary 1917–1945*, Marshall Cavendish (1980), p. 37.
9. Bayerarchiv (BA), 11/3. Vol. 5, 32nd *Aufsichtsrat* meeting (29 September 1939).
10. Six years later at Nuremberg, Ambros was sentenced to eight-year imprisonment for 'slavery and mass murder'.
11. ICHO/RES/0045, 'Properties of Fluoro-tabun: An Analysis of Continuing Research' (February 1945).
12. PRO, WO/42/26, 'Porton Down Chemical Defence Establishment', Correspondence 1939–1945.
13. CIOS Report 31, 'Chemical Warfare Installations in the Munster area' (1945).
14. This figure was the British and American estimate. The Soviet estimate was 250,000 tons, See Jeremy Paxman and Richard Harris, *A Higher Form of Killing: The Secret Story of Chemical Warfare*, New York: Hill & Wang (1982).
15. Greg Goebbel, *A History of Chemical Warfare* 2 (1990), p. 4.
16. The name Trilon 83, and Trilon 146 for sarin, was part of an elaborate deception to keep the Allies in the dark over the nerve gases. Trilon was a common detergent, and as such would arouse no suspicions. Indeed, the British employed similar deceptions, for example, naming the atomic bomb project 'tube-alloys' after a relatively innocuous war material.
17. J. Paxman and R. Harris, *Higher*, op. cit., p. 66.
18. *Ibid.*, p. 67.
19. BA 15/Da.1.1. Umsätze.
20. Interestingly, Degesh was a subsidiary of the German chemical combine, IGFarben.
21. Kim Coleman, 'IG Farbenindustrie AG and Imperial Chemical Industries Limited 1926–1953: Strategies for Growth and Survival', PhD thesis, University of London (2001).
22. BIOS 138, 'Interrogation of German Chemical Warfare Personnel' (1945).
23. *Ibid.*
24. 'Hitler's Deadly Secrets', *The Sunday Times* (22 February 1981).
25. BIOS, Final Report 534, 'The Organisation of the German Chemical Industry and its Development for War Purposes' (7 May 1946).
26. David Irvine, *Hitler's War*, London (1977), p. 633.
27. Imperial War Museum, Speer Papers, FD/4761 (1945).
28. BIOS 542, 'Interrogation of Certain German Personalities Connected with Chemical Warfare' (1945), p. 25.
29. *Trial of Major War Criminals*, 43 Vols, Nuremberg International Tribunal (1947), Vol. 7, pp. 527–528.
30. SIPRI, Vol. 1, p. 135.
31. Omar Bradley, *A Soldiers Story*, New York (1970), p. 237.
32. SIPRI, Vol. 1, p. 297.
33. BA. 4C9.32, Dyestuffs (n.d.).
34. CIOS Report 30 op. cit.
35. Churchill Papers (CHAR) 9/188A–B, Speech-notes-non-House of Commons.
36. Sir John Dill, The *Use of Gas in Home Defence*, HMSO (1940).
37. *Sunday Herald* (9 February 2003).
38. Greg Goebel, op. cit., p. 3.
39. *Sunday Herald*, op. cit.

40. H. Oschsner, *History of German Chemical Warfare in World War II*, Part 2, 'The Military Aspect', Historical Office of the Chief of the [US] Chemical Corps (1949). This quotation is from the review of German chemical warfare activities written in 1948 by General Oschsner, Commander-in-Chief of the German Chemical Corps, 1939–1945. This review is one of the most comprehensive published sources of information on German wartime chemical warfare activities. As source material it is open to criticism that it represents the view of one man only, however closely he was involved in the German chemical warfare programme. Nevertheless, this review was not written for open literature or international circulation. Indeed, it remained a restricted document for thirty years.
41. *Ibid.*
42. G. Alexandrov, 'Lessons of the Past are not to be Forgotten', *International Affairs* (1969), pp. 76–78.
43. W.S. Churchill, *The Hinge of Fate (The Second World War)*, Vol. 4, London (1951).
44. *The Times* (11 May 1942), p. 5.
45. 'Germany and Gas Warfare: Berlin Framing Reply', *The Times* (14 May 1942), p. 3.
46. 'Poland's Gallant Fight Against Odds', *The Times* (6 September 1939), p. 6.
47. 'German Use of Poison Gas', *The Times* (21 November 1939), p. 6.
48. 'German Poison Gas lies', *The Times* (13 October 1939), p. 10.
49. H. Oschsner, op. cit., *History of German Chemical Warfare in World War II*.
50. 'Blow for Russia', *The Times* (11 May 1942), p. 4.
51. Oschsner, op. cit., refers to the potential usefulness of gas for smoking Soviet troops and partisans out of the catacombs in Odessa, but says that orders were issued specifically prohibiting such usage.
52. G. Aklexandrov, 'Lessons of the Past are not to be Forgotten', *International Affairs*, 2 (1969), pp. 76–78.
53. *Sunday Herald*, op. cit.
54. PRO, FO 371, General correspondence.
55. PRO, FO 935, Intelligence Objective Subcommittee, 1933–1945.
56. The Chinese Communists would later carry on such accusations against the Japanese, but the Chinese Communists also made similar accusations against the Americans during the Korean War.
57. Lord Richie-Calder in Steven Rose (ed.), *CBW: London Conference on CBW*, London: George G. Harrap (1968), p. 14.
58. BIOS/JAP/PR/395, 'Intelligence Report on Japanese Chemical Warfare' (March 1946).
59. BIOS, PB 47225, Office of Chief Chemical Officer, GHQ, Tokyo (May 1946).
60. S.P. Lovell, *Of Spies and Stratagems*, New York (1963).
61. F.J. Brown, *Chemical Warfare: A Study in Restraints*, New Jersey: Princeton University Press (1968), pp. 74–75.
62. T.B. Allen and N. Polmar, 'Poisonous Invasion Prelude', *Pittsburgh Post-Gazette* (4 August 1995).
63. US Chemical Warfare Service, *US Chemical Warfare Policy*, Washington DC: Operations Division, War Department General Staff, Strategy and Policy Group (14 June 1945), Draft.
64. *Ibid.*
65. F.J. Brown, op. cit., *Chemical Warfare*.

66. Fritz Haber, Fünf Vorträge (Fifth Lecture), Berlin (1924).
67. D. Irving, *The Destruction of Dresden*, London (1966); H.R. Trevor-Roper, *The Last Days of Hitler*, London (1962 edn); Frederick Taylor, *Dresden Tuesday 13 February 1945*, Bloomsbury (2004); Anton Joachimsthaler, *The Last Days of Hitler: The Legends – The Evidence – The Truth*, Arms & Armour Press (1996).
68. H. Klotz, *Germany's Secret Armaments*, London (1934); H.F. Thuiller, *Gas in the Next War*, London (1939); See also, Bayerarchiv, Leverkusen, IG Farbenindustrie AG, Annual Reports (1934–1944).
69. Oschsner, op. cit.
70. SIPRI, Vol. 1, p. 304.
71. *Daily Telegraph* [Letters to the Editor] (23 July 1968).
72. W.W. Findlay, 'Why Prepare?', *Chemical Industry Journal* (1946), p. 24.
73. K.C. Royall, 'A Tribute to the Chemical Corps', *Chemical Corps Journal* (1947), p. 41.
74. I. King, 'Exit Gas Warfare', *Journal of the International Chemical Industry* (1947), p. 3.

5 The Soviet threat, Korea and Vietnam, 1945–1975

1. LHCMA Nuclear History Database PREM8, Tizard Report (February 1945), www.kcl.ac.uk.lhcma/pro/p-pre08.htm.
2. D.E. Lilienthal, *The Atomic Energy Years 1945–1950*, New York (1964); M. Meselson, 'Why Not Poison?', *Science*, 164 (1969), pp. 413–414.
3. 'Lethal Leftovers', *Scandinavian Times* (November 1970).
4. 'Dangerous Cargo', *The Times* (10 March 1960).
5. SIPRI, Vol. 1, p. 153.
6. 'After 24 Years – Dumped War Gas Hits Holiday Beaches', *Sunday Times* (10 August 1969).
7. *The Times* (17 April 1965).
8. *Sunday Times* (10 August 1969).
9. International implications of dumping poisonous gas and waste into oceans, US House of Representatives, 91st Congress, 1st Session, Washington (May 1969).
10. *Japan Times* (4 October 1970).
11. PRO, WO 193/712, 'Disposal of German Chemical Warfare Stocks' (19 June 1945).
12. *Ibid.*
13. BIOS, Final Report (FR) 542, 'Interrogation of Certain German Personalities Connected with Chemical Warfare' (1946), item no. 8.
14. *The New York Times* (23 February 1938).
15. Quoted in, R.L. Garthoff, *Soviet Strategy in the Nuclear Age*, London (1958), p. 104.
16. *The Penkovsky Papers*, London (1965), p. 153.
17. *CB Weapons Today*, SIPRI, Vol. 1, pp. 173–184.
18. LHMA, 15/5/215 Liddell 1945–1958; General Sir Frederick Alfred Pile, 'This Rocket Can't Miss', *Everybody's Magazine* (6 September 1959).
19. John Pilger, *How Britain Exports Weapons of Mass Destruction*, Znet.org (4 July 2003).

20. Neil Mackay, 'Britain's Chemical Bazaar', *Glasgow Sunday Herald* (9 June 2002).
21. *Ibid.*, p. 2.
22. US Army Chemical Corps, *Summary of Major Events and Problems*, FY58, Army Chemical Center, Md., Historical Center (March 1959), p. 100.
23. Quoted in Greg Goebel, *A History of Chemical Warfare* 2, www.vectorsite.net/gas.html, p. 6.
24. M. Stubbs, 'A Power for Peace', *Armed Forces Chemical Journal*, 13 (1959), pp. 8–9.
25. www.ddh.nl.
26. Seymour Hersh, *Chemical and Biological Warfare: America's Hidden Arsenal*, Bobbs-Merrill (1968).
27. A.C. McAuliffe, 'Korea and the Chemical Corps', *Ordnance*, 35 (1951), p. 284.
28. E.R. Baker, 'Chemical Warfare in Korea', *Armed Forces Chemical Journal*, 4 (1951), p. 3.
29. E.R. Baker, 'Gas Warfare in Korea', *Armed Forces Chemical Journal*, 4(4) (1951), pp. 54–56.
30. *Washington Times* (15 August 1992).
31. The 38th parallel roughly marked the line between the capitalist South Korea and the communist North Korea.
32. 'The Great Chemical Weapons Hoax', www.int.org/chemwar0503a.html, p. 13.
33. PBS, TV Programme, *Race for the Superbomb* (January 1999).
34. International Committee of the Red Cross, 'Le Conflict de Corée', Documents, Vol. 11, Geneva (1952), pp. 79–83.
35. United Nations Security Council, S/2684/Add 1 (30 June 1952).
36. E.F. Bullene, 'The Needs of the Army', *Armed Forces Chemical Journal*, 6 (1952), p. 8.
37. A. Harrigan, 'The Case for Gas Warfare', *Armed Forces Chemical Journal*, 17 (1963), p. 12.
38. *Ibid.*, p. 14.
39. 'Mustard Gas Use Suggested', *Washington Post* (18 October 1966), A2.
40. *Scientist and Citizen* (August–September 1967), p. 152.
41. G.W. Merck, 'Report to the Secretary of War', *Military Surgeon*, 98 (1946), pp. 237–242.
42. H.W. Baldwin, *Great Mistakes of the War*, New York (1950).
43. Chemical Warfare Hearing before the subcommittee on National Security Policy and Scientific Developments, US House of Representatives, 91st Congress, Washington (November–December 1969).
44. Congressional Record, S.14240–80 (26 August 1970).
45. A. McConnell, 'Mission: Ranch Hand', *Air University Review*, 21 (1970), pp. 12–15.
46. *The New York Times* (10 September 1966).
47. *St Louis Post-Dispatch* (11 July 1967).
48. *The Financial Times* (18 July 1967).
49. *The New York Times* (10 September 1966).
50. *The Merck Index of Chemicals and Drugs*, 7th edn, Rahway, New Jersey: Merck & Co. (1960), p. 184.
51. M.K. Kahn, *CBW in Use – Vietnam*, London: George G. Harrap & Co. Ltd (1968), p. 90.
52. *Baltimore Sun* (15 January 1967).
53. L.F. Fieser, 'Napalm', *Industrial and Engineering Chemistry*, 38 (1946), pp. 768–773.

54. B.E. Kleber and D. Birdsell, *The Chemical Warfare Service: Chemicals in Combat*, Washington DC (1966), p. 534.
55. The programme is presented in great detail in L.F. Feiser, *The Scientific Method: A Personal Account of Unusual Projects in War and Peace*, New York: Reinhold (1964).
56. F.J. Sanborn, 'Fire Protection Lessons of the Japanese Attacks', in H. Bond (ed.) *Fire and the Air War*, Boston (1946), pp. 169–174.
57. 'Napalm', *Armed Forces Chemical Journal*, 7 (July 1953), p. 8.
58. Interview with Tom Jackson (1 November 2003).
59. *Ibid.*
60. E.W. Hollingsworth, 'Use of Thickened Gasoline in Warfare', *Armed Forces Chemical Journal*, 4 (June 1951), pp. 26–32.
61. *The Times* (11 December 1967).
62. 'Use of Lethal Gases Charged', *The New York Times* (6 April 1965).
63. Commission for the Investigation of the American Imperialists War Crimes in Vietnam (October 1966). Source located at the Vietnamese Embassy, London.
64. *Time* (15 June 1998).
65. *Ibid.*
66. *Ibid.*
67. SOG units did not 'officially' exist. According to *Time* they conducted 'black operations' with unconventional weapons and unusual targets. According to SOG veterans they had no rules of engagement: anything was permissible as long as it was deniable. Their motto: 'Kill them all, and let God sort it out'.
68. Mike Hagan quoted in *Time* (15 June 1998).
69. *Ibid.*
70. www.copvcia.com.
71. See, for example, Monika Jensen-Stevenson, *Kiss the Boys Goodbye*, Dutton (1990).
72. www.teamhouse.tni.net/Tailwind/ruppert.htm.
73. Reprinted by permission, Michael C. Ruppert and From the Wilderness at www.copvcia.com.
74. Quoted in Kahn, op. cit., p. 92.
75. PRO, Porton Technical Paper, No. 651 (1970).
76. *The New York Times* (2 February 1970).
77. *The New York Times* (25 March 1965).
78. *The New York Times* (26 March 1965).
79. S.M. Hersh, *Chemical and Biological Warfare: America's Hidden Arsenal*, New York (1968).
80. *Ibid.*
81. *Time* (2 April 1965).
82. See, for example, Associated Press (10 January 1966; 12 January 1966).
83. *The New York Times* (13 January 1966).
84. *The New York Times* (25 March 1965).
85. *The New York Times* (24 March 1965).
86. Wil D. Verweg, *Riot Control Agents and Herbicides in War*, A.W. Sijthoff International Publishing Company BV (1977), p. 242.
87. United Nations Document (UND), 455, 456.
88. UND, 460.

89. Revelation 16, verses 16–18.
90. Erika Cheetham, *The Prophecies of Nostradamus*, The Bath Press (1974), p. 353.

6   The Middle East, Afghanistan, Bosnia and the Gulf

1. *Daily Telegraph* (8 July 1963).
2. M. Meselson, 'The Yemen', in Steven Rose (ed.), *Chemical and Biological Warfare*, London: Harrap (1968), p. 99.
3. *Ibid.*
4. *Washington Post* (6 June 1967).
5. See, for example, *Daily Telegraph, The Times, The New York Times, Washington Post* (6–8 June 1967).
6. www.vnh.org/chapter2.htm.
7. Red Cross Statement, quoted in Meselson, op. cit., p. 100.
8. D.H. Schmidt, *Yemen: The Unknown War*, London (1968).
9. *Daily Telegraph* (21 July 1967).
10. D.M. van Rosmalen, 'Yemen: A Testing Ground for Poison Gas', *Elseviers' Weekblad* (25 November 1967).
11. S.M. Hersh, *Chemical and Biological Warfare: America's Hidden Arsenal*, New York: Bobbs-Merrill (1968), pp. 286–287.
12. US Army Chemical Corps, *Summary of Major Events and Problems*, FY (1961), No. 62, 99, pp. 9–20.
13. SIPRI, Vol. 11, p. 193.
14. www.vnh.org/htm.
15. Reducing the terror of war, *Commanders-Digest* (20 December 1969), p. 4.
16. Karel Knip, Biologie in Ness Ziona, *NRC Hanelsband* (27 February 1999).
17. See, for example, SIPRI, *The Problems of Chemical and Biological Warfare*, Stockholm (1971), Vol. 1, pp. 75–78; Vol. 11, pp. 35–36.
18. ARRADCOM: Fiscal Year 1978, pp. 3–4, quoted on www.vnh.org/htm.
19. www.munition.gov.ru/eng/oldobjl.htm (Soviet Munitions Agency).
20. R.L. Wagner and T.S. Gold, 'Why We Can't Avoid Developing Chemical Weapons', *Defense* (1982), p. 3; *Fort McClellan*, National Military Publications (1983), pamphlet.
21. Department of Defense, Appropriation Authorization Act, 95–79 (30 July 1977).
22. www.vectorsite.net/twgas2.html.
23. A.M. Haig Jr, *Chemical Warfare in South East Asia and Afghanistan*, Report to Congress, Washington DC: Department of State (1982), Special Report No. 98.
24. J.J. Collins, 'The Soviet Military Experience in Afghanistan', *Military Review* (1985), p. 27.
25. SIPRI, Chemical Warfare Factsheet (1984).
26. *Ibid.*
27. *Ibid.*
28. SIPRI, op. cit., Vol. 1, p. 74.
29. *Ibid.*
30. *Ibid.*
31. Quoted in P. Dunn, *Chemical Aspects of The Gulf War: Investigations by the United Nations*, Ascot Vale, Australia (1987).

32. R.M. Coik-Deegan *et al.*, *Winds of Death: Iraqs Use of Poison Gas Against Its Kurdish Population*, Report of a Medical mission to Turkish Kurdistan by Physicians for Human Rights, Boston: Massachusetts (1989).
33. *Ibid.*
34. www.hrw.org/reports/bosni98-02.htm.
35. Interviews conducted by Human Rights Watch, February–August 1996, www.hrw.org/reports98.htm.
36. JNA, *Speciajalne Rucne Bombe M-79*, Belgrade (1988). Translated by Human Rights Watch (1999).
37. David Rohde, *Endgame: The Betrayal and Fall of Srebrenica, Europe's Worst Massacre Since World War II*, New York (1997), pp. 266–273.
38. www.hrw.org/reports98/bosniacw/htm.
39. *Ibid.*, Part IV, p. 3.
40. A.T. Herbig, 'Nerve Agents and Their Psychological Effects', *US Army Chemical Review* (1990), pp. 9–13.
41. www.cnn/news/archive.htm.
42. S.S. Johnson, 'Cheney Fears Chemical Attack Possible', *USA Today* (28 January 1991).
43. *Ibid.*
44. www.mod.uk/issues/gulfwar/info/ukchemical.htm, p. 2.
45. *Ibid.*, p. 3.
46. Jonathan B. Tucker, *Non-Proliferation Review* (Spring–Summer 1997), p. 114.
47. *Ibid.*, p. 114.
48. *Newsweek* (4 March 1991).
49. Tucker, op. cit., p. 115.
50. 'Gulf War Veterans Illness – Dealing with the Uncertainties', POST, Parliamentary Bookshop (1997).
51. *Ibid.*
52. BBC2, Horizon, Gulf War Jigsaw (14 May 1998).
53. POST Report Summary (December 1997); www.parliament.uk/post/home/htm.
54. BBC2, Horizon, op. cit.
55. *The Times* (23 October 1996).
56. David Guyatt, 'Biological Black Magic', *Nexus Magazine*, Vol. 4 (August–September 1997).
57. D. Bernstein, 'Gulf War Syndrome Cover-Up', *Covert Action Quarterly* (Winter 1992/1993).
58. BBC2, Horizon, op. cit.
59. http://osins.sunderland.ac.uk/autism/hooper2000.htm, p. 2.
60. Defence Committee, 7th Report Gulf Veterans Illness, HMSO (April 2000).
61. BBC2, Horizon, op. cit.
62. RAND Report, US Department of Defense (14 October 1999).
63. UNSCOM, 1991, 1992, 1995.
64. UNSCOM, 1995.
65. UNSCOM, 1991, 1992, 1995.
66. UNSCOM, 1995.
67. Resolutions 687 (1991) and 699 (1992) included disarmament clauses which demanded Iraq declare its programmes of weapons of mass destruction and long-range missiles, verified the declarations through UNSCOM and allowed

supervision of the destruction or elimination of prescribed programmes and items.

68. Parliamentary Office of Science and Technology, POSTnote 111 (February 1998), p. 1.
69. www.cia.gov/cia/reports/iraq_wmd/Iraq_Oct_2002.htm, p. 2.
70. *The Guardian* (6 March 2003).
71. www.gwu.edu/~nsarchiv/NSAEBB, p. 2.
72. www.un.org/Depts/Unmovic/Bx27.htm, p. 4.
73. *The Guardian* (18 March 2003).
74. Hans Blix, *Disarming Iraq*, Pantheon Books (2004), Chapter 1, 'Moments of Truth'.
75. CIA, *Iraq's Weapons of Mass Destruction Programs* (October 2002). Full text available at www.cia.gov.
76. Prime Minister Tony Blair's Introduction to *Iraq's Weapons of Mass Destruction: The Assessment of the British Government* (September 2002), p. 3. Full text available at www.pm.gov.uk.
77. *Toronto Sun* (21 April 2003).
78. *The Butler Report*, HMSO (2004), p. 213.
79. Op. cit., *Iraq's Weapons*, p. 54.
80. *The Independent* (7 July 2004).

7 Chemical terrorism

1. Donna Shalala, 'How Prepared are We?', *Emerging Diseases*, 5, No. 4 (July/August 1999).
2. www.fas.org/irp/threat/an253stc.htm.
3. For a succinct overview of the roots of terrorism, see Walter Lacqueur, *The New Terrorism: Fanaticism and the Arms of Mass Destruction*, Oxford University Press (1999), pp. 8–48.
4. Brian M. Jenkins, 'Understanding the Link Between Motives and Methods', in Brad Roberts (ed.), *Terrorism with Chemical and Biological Weapons*, Arlington VA: The Chemical and Biological Arms Control Institute (1997), p. 45.
5. Brian M. Jenkins, 'Will Terrorists Go Nuclear?', *Orbis*, No. 3 (Autumn 1985), p. 511.
6. See, for example, Robert Mullen, 'Mass Destruction and Terrorism', *Journal of International Affairs*, No. 1 (Spring/Summer 1978), pp. 63–89; Jenkins, op. cit., *Understanding*, pp. 507–515.
7. These trends are explained briefly in 'The New Terrorism: Coming Soon to a City Near You'. *The Economist* (15 August 1988).
8. See, *The New York Times* (18 December 1996) and 'Peru Troops Rescue Hostages: Rebels Slain as Stand Off Ends', *The New York Times* (23 April 1997).
9. *The New York Times* (13 December 1995).
10. www.geocities.com/rollofhonor32/downingst.html.
11. *Ibid.*
12. For more detail on these cult stories, see Tom Matthews, 'The Cult of Death', *Newsweek* (4 December 1978), pp. 38–43; Barry Bearak, 'Eyes on Glory: Pied

Pipers of Heaven's Gate', *The New York Times* (28 April 1997); also *Washington Post* (19 March 2000; 31 March 2000).

13. Bruce Hoffman, *Inside Terrorism*, New York: Columbia University Press (1998), pp. 93–94.

14. Boaz Ganor, 'Israel's Counter-Terrorism Strategy', PhD in progress, Hebrew University (2003).

15. *Ibid.*

16. *Ibid.*

17. See, Ashton B. Carter *et al.*, *Catastrophic Terrorism: Elements of a National Policy*, Stanford University, Cambridge, MA: Harvard University (1998); Ashton B. Carter, 'Catastrophic Terrorism: Tackling the New Danger', *Foreign Affairs*, No. 6 (November/December 1998).

18. Amy E. Smithson, *Ataxia: The Chemical and Biological Threat and the US Response*, Washington DC: Henry L. Stimson Center (1999), p. 73.

19. *Ibid.*, p. 74.

20. David E. Kaplan and Andrew Marshall, *The Cult at the End of the World*, New York: Crown (1996), pp. 107–108.

21. *Chemical and Engineering News* (31 August 1998), p. 7.

22. The quantity of sarin made in October 1993 was 20 gms; 1 kg in November 1993; 5 kg in December 1993; 20 kg between February and March 1994. Anthony Tu, 'Chemistry and Toxicology of Nerve Gas Incidents in Japan in 1994 and 1995', paper presented at the Third Congress of Toxicology in Developing countries (1 September 1996), Cairo.

23. D.W. Brackett, *Holy Terror: Armageddon in Tokyo*, New York: Wetherhill (1996), p. 116.

24. Kaplan and Marshall, op. cit., *The Cult*, pp. 144–146.

25. www.fas.org/irp/threat/an253stc.htm.

26. *The Washington Post* (February 2001), article reprinted at www.washingtonpost.com/ac2/wp-dyn.htm.

27. The most detailed account of the attack can be found in Brackett, *Holy Terror*, op. cit., pp. 132–134.

28. Brackett, *Holy Terror*, op. cit., pp. 134–140.

29. F.R. Sidell, 'Chemical Agent Terrorism', *Annals of Emergency Medicine*, 28, No. 2 (August 1996), p. 224.

30. www.vectorsite/gas/tw5.html.

31. Cases in point include, *The Rock* (1996); *Mission Impossible*, 2 (2000).

32. President William J. Clinton, 'Interview of the President by the New York Times', Washington DC: White House Office of the Press Secretary (23 January 1999); Richard Preston, *The Cobra Event*, Random House (1998).

33. This survey is available in its entirety at www.ccfr.org/publications/opinion/opinion.html.

34. David C. Rapport, 'Terrorism and Weapons of the Apocalypse', *National Security Studies Quarterly*, No. 3 (Summer 1999), p. 60.

35. www.net/gas/tw5.html.

36. Stephen Sloan, 'If There is a "Fog of War", There is Probably a More Dense "Smog of Terrorism"', in *Terrorism: National Security Policy and the Home Front*, Strategic Studies Institute, United States Army War College (1995), p. 51.

37. In 1973, the Symbionese Liberation Army killed a school superintendent in Oakland, California, with a bullet tipped with cyanide. Jonathan B. Tucker

and Amy Sands, 'An Unlikely Threat', *Bulletin of Atomic Scientists*, No. 4 (July/August 1999), p. 46.

38. Peter Hirschberg, 'Inside the Low White Buildings', *Jerusalem Report* (21 December 1998), p. 18.

39. Harms van den Berg, 'The Fatal Fallout from El Al Flight 1862', *Jerusalem Report* (21 December 1998), pp. 16–21. For a comprehensive review of Israel's chemical and biological weapons programme, see Avner Cohen, 'Israel and Chemical/Biological Weapons: History, Deterrence and Arms Control', *The Non-Proliferation Review*, Fall–Winter (2001), pp. 27–53.

40. Jenkins, 'Understanding the Link', op. cit., p. 49.

41. Brackett, *Holy Terror*, op. cit., p. 6.

42. Hoffman, *Inside Terrorism*, op. cit., p. 205.

43. Jeffrey D. Simon, *Toxic Terror: Assessing the Terrorist Use of Chemical and Biological Weapons*, Cambridge: MIT Press (2000), pp. 17, 73–74.

44. Lacqueur, *New Terrorism*, op. cit., pp. 265–267.

45. Tucker and Sands, 'An Unlikely Threat', op. cit., p. 51.

46. Amy E. Smithson, *Ataxia*, op. cit., p. 30; Amy E. Smithson, *Separating Fact from Fiction: The Australia Group and the Chemical Weapons Convention*, Washington DC: Henry L. Stimson Center (1997).

47. Office of Technical Assessment, *Technologies Underlying Weapons of Mass Destruction*, Washington DC: US Government Printing Office (1993), p. 27.

48. SIPRI, *Rise of CB Weapons*, op. cit., Vol. 1, p. 76.

49. CIA, *The Chemical and Biological Weapons Threat* (1996), 15.

50. For a detailed discussion see, SIPRI, *Rise of CB Weapons*, op. cit., Vol. 2, pp. 72–90.

51. Indeed, following 9/11 crop dusters were not allowed to operate in air space near US military installations.

52. www.rand.org/terrorism.htm.

53. *Ibid.*

54. Amy E. Smithson, *Ataxia*, op. cit., p. 57.

55. *Ibid.*

56. www.rand.org., op. cit.

57. For a detailed analysis of the Monterey database statistics see, Smithson, *Ataxia*, op. cit., pp. 61–65.

58. *Ibid.*, p. 62.

59. *Associated Press* (9 July 1998).

8 Controlling chemical weapons

1. Interview with Julian Perry Robinson, SPRU, University of Sussex (5 July 2004).

2. See, for example, the Indian epics, *Ramayana* and *Mahabharata*. Also, later Chinese and Greek sources.

3. Hague Declaration 1899, Article IV(2).

4. 'Entry in Force Factsheet', SIPRI (no date).

5. Perry Robinson, op. cit.

6. See, SIPRI, *The Problems of Chemical and Biological Warfare*, 6 Vols, Stockholm (1971–1975).

7. J.P. Perry Robinson *et al.*, 'The Chemical Weapons Convention: The Success of Chemical Disarmament Negotiations', *SIPRI Yearbook 1993: Armaments,*

Disarmament and International Security, Oxford University Press (1993), pp. 705–734.

8. CWC, Article VI(a).
9. CWC, Article X, paragraph 2. States were allowed to produce 1 ton annually of chemical agents for research, medical and pharmaceutical or protective purposes. Monitoring of these activities is detailed in CWC Verification Annex, Part IV, Sections A–E.
10. Declaration requirements and destruction timetables can be found in CWC, Article III.
11. CWC, Article VI.
12. CWC, Article VI; Verification Annex, Parts VI–IX; Annex on Chemicals.
13. CWC, Verification Annex, Part II, paragraphs 27, 45 and 51–58.
14. United States Department of Defense, *Proliferation: Threat and Response*, Washington DC: United States Government Printing Office (1997).
15. www.sipri.se/cbw/research/ssf—eif.html.
16. Note that the production facility declared by Japan was Satyam 7, located at Kamikuishiki, formerly used by the religious cult Aum Shinrikyo which released the nerve gas sarin on the Tokyo underground system on 20 March 1995.
17. Perry Robinson, op. cit.
18. Fact sheet, 'US–Soviet Memorandum of Understanding on Chemical Weapons', Washington DC: White House (23 September 1989).
19. Amy E. Smithson and Maureen Lenihan, 'The Destruction of Weapons Under the Chemical Weapons Convention', *Science and Global Security*, 6 (1996), p. 93; Jonathan B. Tucker, 'Russia's Plan for Chemical Weapons Destruction', *Arms Control Today* (July–August 2001).
20. State parties to CWC as at 27 May 2004.
21. CWC, Verification Annex, Part VII, paragraphs 31–32.
22. Challenge inspections are dealt with in Article IX, CWC.
23. Amy E. Smithson, 'Rudderless: The Chemical Weapons Convention at 11/2', Report No. 25, Washington DC: Henry L. Stimson Center (1998), p. 66.
24. Tests Show Nerve Gas in Iraqi Warheads: Finding Contradicts Baghdad Claims, *Washington Post* (23 June 1998), A1.
25. 'US Offers More Details on Attack in the Sudan', *The New York Times* (24 August 1998), A1.

Bibliography

Primary sources

Britain
Records located at Churchill College, Cambridge
Churchill Papers (CHAR)
CHAR 20/23/13–82, Papers relating to Gas Attack (1919–1920).
CHAR 20/16/1–3, Use of Gas in Iraq (1919–1920).
CHAR 9/188 A–B, speech-notes, non-House of Commons (1941–1945).

Records located at Imperial Chemical Industries Head Office (ICHO)
ICHO/Directors Papers (DIR)/1964, KSK Production (April 1927).
ICHO/DIR/0019, History of Imperial Chemical Industries Dyestuffs Division, typed manuscript (1942).
ICHO/DIR/0365, Government Construction Contracts (31 December 1943).
ICHO/DIR/0045, Properties of Fluoro-tabun. Analysis of continuing research (February 1945).

Records located at the Imperial War Museum, London
British Intelligence Objectives Subcommittee (BIOS)
BIOS 714, The Development of New Insecticides and Chemical Warfare Agents (no date).
BIOS 542, Interrogation of Certain German Chemical Warfare Personnel (1946).
BIOS 138, Interrogation of German Chemical Warfare Personnel (1945).
BIOS Final Report 534, The Organisation of the German Chemical Industry and its Development for War Purposes (7 May 1946).

Nuremberg Industrialists Documents (NI)
NI-6788, Affidavit by Otto Ambros (1 May 1947).

Speer Papers (FD)
FD/4761, Patents (1945).

Records located at the Liddell Hart Military Archives, King's College (LHCMA)
LHCMA, Nuclear History Database, PREM8, Tizard Report (February 1945).
LHCMA 15/5/215, Liddell (1945–1958).

Foulkes Papers (FOULKES)
FOULKES 6/37, Establishment of Central Laboratory, Helfaut (1915).
FOULKES 6/6, Papers relating to the Battle of Loos on 25 September 1915 and Sir John French's Dispatch (15 October 1915).
FOULKES 6/5, Papers relating to Livens Projector (1915–1921).
FOULKES 6/45, Official evidence compiled from Germany of the results of British gas attacks (1918–1920).
FOULKES 6/19, Papers relating to filter fans and blowers for dugouts (1917).

FOULKES 6/1–81, Papers relating to the introduction and use of gas warfare (1915–1919).

Records located in the Public Record Office (PRO) now National Archives (NA)

PRO, Air Ministry Papers (AVIA)
PRO, AVIA 10/338, War Cabinet Meetings (1914–1918).

PRO, Foreign Office Papers (FO)
PRO, FO 371, General Correspondence (1933–1939).
PRO, FO 935, Intelligence Objective Subcommittee Reports (1933–1945).

PRO, Ministry of Munitions Series (MUN)
PRO, MUN 4/2735, Chemicals and Their Use by British and German Armies (10 August 1917).
PRO, MUN 4/1684, Gun Ammunition Filling (2 May 1918).

PRO, War Office Series (WO)
PRO, WO 15/8/815, Chlorine Gas (June 1915).
PRO, WO 32/5169, Operations: General Code 46 (B) (June 1915).
PRO, WO 32/5173, Engineers: Code 14 (E), Formation of Special Gas Companies (1915).
PRO, WO 30/57/80, SUPP 10/92, Reports on the Use of Poison Gas (1918).
PRO, WO 32/595, Casualties Caused by Poison Gas in British Forces 1915–1918 (1918–1919).
PRO, WO 32/5176, Report on Gas Attack by Germany (1915).
PRO, WO 106/148, Reports on Mustard Gas (April 1917 to September 1919).
PRO, WO 32/5177, Laws of War: Code 55/E, Threat of New Poison Gas Shells (1918–1919).
PRO, WO 142/71–72, Chemical Warfare Committee 77–112 (27 October 1917 to 28 March 1918).
PRO, WO 118/114, Chemical Warfare – Germany (1918–1922).
PRO, WO 33/1014, 2nd Annual Report of the Chemical Defence Establishment, Porton Down (1922).
PRO, WO 33/1049, 4th Annual Report of the Chemical Defence Establishment, Porton Down (1924).
PRO, WO 142/98–99, DES/1/1–80; DES/2/81–189; DES/4/281–390; DES/5/391–489, Mustard Gas Files (1917–1919).
PRO, WO 208/3025, German Chemical Warfare Material (1945).
PRO, WO 33/1484, Porton Down – Annual Report (1937).
PRO, WO 33/078–1568, Porton Down Annual Reports (1922–1939).
PRO, WO 193/712, Disposal of German Chemical Warfare Stocks (19 June 1945).
PRO, WO 195/9236, Porton Report No. 2747, Preliminary Report on the Potential Value of Nerve Gases as Chemical Warfare Agents (18 January 1947).

Porton Down Technical Papers (TP)
PRO, TP 651 (1970).

Other selected documents
Nuclear Database, PREM 8, Tizard Report (February 1945).

Germany

Records located at the Bayerarchiv (Bayer Archives), Leverkusen (BA)
BA 113, Vol. 5, 32nd Aufsichtsrat (29 September 1939).
BA 15/Da.1.1. Umsätze.

BA 4 C9/32, Dyestuffs (no date).
IG Farbenindustrie AG, Annual Reports (1934–1944).

Records located at the East German State Military Archive (Militärarchiv der Deutschen Demokratischen Republik) (MA)
MA W31.60/20.

United States

Records located at Edgewood Arsenal Archives
Status Summary of the Relative Values of AC, CK and CG as bomb fillings, Staff Report No. 1 (10 July 1944).

Records located at National Archives (NA), Washington
Office of Chief Chemical Officer, GHQ, AFPAC, Intelligence Report on Japanese Chemical Warfare (May 1946).
Office of Chief Chemical Officer, New Gas Detector Kit, Information (12 September 1944).
Record Group (RG) 281, Miscellaneous World War II.

Records located at United States Chemical Warfare Service, History Division
US Chemical Warfare Service, Operations Division, War General staff, Strategy and Policy Group (14 June 1945).

Official publications (all countries)

CIA, *Iraq's Weapons of Mass Destruction Programs* (October 2002).
Haig, A.M. Jr, Special Report No. 98, *Chemical Warfare in South East Asia and Afghanistan*, Report to Congress, Washington DC: Department of State (1982).
H.M. Defence Committee, 7th Report, *Gulf War Veterans Illness*, HMSO (April 2000).
H.M. Government, Parliamentary Office of Science and Technology, POSTnote 111 (February 1998).
H.M. Government, *Iraq's Weapons of Mass Destruction: The Assessment of the British Government* (September 2002).
RAND Report, United States Department of Defense (14 October 1999).
The Butler Report, HMSO (July 2004).
UNSCOM Reports (1991, 1992, 1995).
US Government, Office of Technical Assessment, *Technologies Underlying WMDs* (1993).

Conventions, treaties, acts of parliament

Chemical Warfare Convention (1995).
Geneva Protocol (1925).
Hague Declaration (1899).
Potsdam Agreement (1945).
St Petersburg Declaration (1868).
Strasbourg Agreement (27 August 1675).
Treaty of St Germaine (1920).
Treaty of Trianon (1920).
Treaty of Versailles (1919).
United States Department of Defense, Appropriation Authorization Act 95–79 (30 July 1977).

Miscellaneous

International Committee of the Red Cross, 'Le Conflict de Corée', Documents, Vol. 11, Geneva (1952).
United Nations Documents (UND), 455, 456, 460.
United Nations Security Council, S/2684/Add 1 (30 June 1952).

Secondary sources

Journals

Alexandrov, G., 'Lessons of the Past are not to be Forgotten', *International Affairs*, 2 (1969).
Anon., 'T-Stoff', *Army and Navy Journal* (May 1915).
——, 'Chemical Warfare Bombing Tests on Warships', *Chemical Warfare*, 6(10) (1921).
——, 'Napalm', *Armed Forces Journal* (July 1953).
——, 'BC Stridsmedel', *Forsvarets Forkninganstalt*, No. 2 (1964).
——, 'Reducing the Terror of War', *Commander's Digest* (20 December 1969).
Baker, E.R., 'Chemical Warfare in Korea', *Armed Forces Chemical Journal* (1951).
Brett, H.H., 'Chemicals and Aircraft', *Chemical Warfare Bulletin*, 23 (1936).
Bullene, E.F., 'The Needs of the Army', *Armed Forces Chemical Journal* (1952).
Carter, Ashton B., 'Tackling the New Danger', *Foreign Affairs*, No. 6 (November–December 1998).
Chemical and Engineering News (31 August 1998).
Chemical Trade Journal (1919).
Cohen, Avner, 'Israel and Chemical and Biological Weapons: History, Deterrence and Arms Control', *Non-Proliferation Review* (Fall–Winter 2001).
Collins, J.J., 'The Soviet Military Experience in Afghanistan', *Military Review* (1985).
Duffield, M., 'Ethiopia: The Unconquered Lion of Africa', *Command Magazine*, 4 (1990).
Fieser, L.F., 'Napalm', *Industrial and Engineering Chemistry*, 38 (1946).
Findlay, W.W., 'Why Prepare?', *Chemistry Industry Journal* (1946).
Fries, A.A., 'Sixteen Reasons Why the Chemical Warfare Service Must be a Separate Department of the Army', *Chemical Warfare*, 2(1) (1920).
——, 'Chemical Warfare: Past and Future', *Chemical Warfare*, 5(1) (1920).
Fuller, J.C., 'Chemicals in Ethiopia', *Chemical Warfare Bulletin*, 22 (1936).
Galgarraga, H.E., 'The Evolution of the Protective Mask for Military Purposes: Inception tö World War I', *Greenwich Journal of Science and Technology*, 2(1) (2001).
Ghosh, R. and Newman, J.F., 'A New Group of Organophosphorous Pesticides', *Chemistry and Industry* (29 January 1955).
Harrigan, A., 'The Case for Gas Warfare', *Armed Forces Chemical Journal* (1963).
Herbig, A.T., 'Nerve Agents and Their Psychological Effects', *United States Army Review* (1990).
Hirschberg, Peter, 'Inside the Low White Buildings', *Jerusalem Report* (21 December 1998).
Hollingsworth, E.W., 'The Use of Thickened Gasoline in Warfare', *Armed Forces Chemical Journal*, 4 (June 1951).

Jenkins, Brian M., 'Will Terrorists Go Nuclear?', *Orbis*, No. 3 (Autumn 1985).

King, I., 'Exit Gas Warfare', *Journal of the International Chemical Industry* (1947).

Knipp, Karel, 'Biologie in Ness Ziona', *NRC Hanelsband* (27 February 1999).

Liddell-Hart, B., 'The Abyssinian War', *Ordnance Magazine* (1937).

McAuliffe, A.C., 'Korea and the Chemical Corps', *Ordnance Magazine* (1951).

McConnell, A., 'Mission Ranch-Hand', *Air University Review*, 21 (1970).

Merck, G.W., 'Report to the Secretary of War', *Military Surgeon*, 98 (1946).

Meselson, M., 'Why not Poison?', *Science*, 164 (1969).

Mullen, Robert, 'Mass Destruction and Terrorism', *Journal of International Affairs*, No. 1 (Spring–Summer 1978).

Nature (12 January 1922).

Pile, General Sir Frederick Alfred, 'This Rocket Can't Miss', *Everybody's Magazine* (6 September 1959).

Rapport, David, 'Terrorism and Weapons of the Apocalypse', *National Security Studies Quarterly*, No. 3 (Summer 1999).

Regan, Geoffrey, 'Blue on Blue', *The Historian* (Summer 2003).

Royall, K.C., 'A Tribute to the Chemical Corps', *Chemical Corps Journal* (1947).

Sartori, M., 'New Developments in the Chemistry of War Gases', *Chemical Review*, 48 (1951).

Scientist and Citizen, Editorial (August–September 1967).

Shalala, Donna, 'How Prepared are We?', *Emerging Diseases*, 5 (July–August 1999).

Sidell, F.R., 'Chemical Agent Terrorism', *Annals of Emergency Medicine*, 28(2) (August 1996).

Sloan, Stephen, 'If There is a "Fog of War", There is Probably a Dense "Smog of Terrorism"', *National Security Policy Review* (1995).

Stockholm International Peace Research Institute (SIPRI), *Chemical Warfare Factsheet* (1984).

Stubbs, M., 'A Power for Peace', *Armed Forces Chemical Journal*, 4 (1951).

Tucker, Jonathan, *Non-Proliferation Review* (Spring–Summer 1997).

Wagner, R.L. and Gold, T.S., 'Why We Can't Avoid Developing Chemical Weapons', *Defense* (1982).

Wood, J.R., 'Chemical Warfare – A Chemical and Toxicological Review', *American Journal of Public Health*, 34 (1946).

Books

5250th Technical Intelligence Company, *The Use of Poison Gas by Imperial Japanese Army in China 1937–1945*, Tokyo: TIC (1946).

Angell, N., *The Menace to Our National Defence*, London: Arnold (1934).

Ashton, E. *et al.*, *Catastrophic Terrorism: Elements of a National Policy*, Cambridge, MA: Stanford University (1998).

Baldwin, H.W., *Great Mistakes of the War*, New York: Duell & Sloane (1950).

Barrie, A., *War Underground*, New York: Staplehurst-Spellmount (2000).

Berton, Pierre, *Vimy*, Toronto: McClelland & Stewart (1986).

Blix, Hans, *Disarming Iraq*, New York: Pantheon (2004).

Bond, H. (ed.), *Fire and the Air War*, Boston: n.p. (1946).

Brackett, D.W., *Holy Terror: Armageddon in Tokyo*, New York: Wetherill (1996).

Bradley, Omar, *A Soldiers Story*, New York: Duell & Sloane (1970).

Brandon, Piers, *The Dark Valley*, Cambridge: Cambridge University Press (1938).

Brophy, L.P. and Fisher, G.J.B., *The Chemical Warfare Service: From Lab to Field*, Washington DC: Office of the Chief of Military History (1959).

Brown, F.J., *Chemical Warfare: A Study in Restraints*, New Jersey: Princeton University Press (1968).

Brown, Malcolm, *Tommy Goes to War*, London: J.M. Dent & Sons (1978).

Carter, C.B., *Porton Down: 75 Years of Chemical and Biological Research*, London: HMSO (1992).

Cheetham, Erika, *The Prophecies of Nostradamus*, London: Bath Press (1974).

Churchill, W.S., *The Hinge of Fate (The Second World War)*, Vol. 4, London: Edward Arnold (1951).

Clark, D.K., *Effectiveness of Toxic Weapons in the Italian–Ethiopian War*, Bethesda: Operations Research Office (1959).

Coik-Doegan, R.M., Howard Hu and Asfandiar Shukri, *Winds of Death: Iraq's Use of Poison Gas Against Its Kurdish Population*, Report of a Medical Mission to Turkish Kurdistan by Physicians for Human Rights, Boston: John Hopkins University (1989).

Dill, John, *The Use of Gas in Home Defence*, London: HMSO (1940).

Dundonald, Lt. Gen. The 12th Earl, *My Army Life*, London: Arnold (1926).

Dunn, P., *Chemical Aspects of the Gulf War: Investigations by the United Nations*, Australia: Ascot Vale (1987).

Edmonds, Sir J.E., *Military Operations: France and Belgium*, Vol. iii, London: Ascot Vale (1927).

Feiser, L.F., *The Scientific Method: A Personal Account of Unusual Projects in War and Peace*, New York: Reinhold (1964).

Foulkes, C.H., *Gas! The Story of the Special Brigade*, Edinburgh: William Blackwood and Sons (1936).

Gartoff, R.L., *Soviet Strategy in the Nuclear Age*, London: Arnold (1958).

Graves, Robert, *Goodbye to all That*, New York: Doubleday (1957).

Haber, F., *Fünf Vorträge*, Berlin: Frei Universität (1924).

Haber, L.F., *The Poisonous Cloud: Chemical Warfare in the First World War*, Oxford: Clarendon Press (1986).

Haldane, J.C.S., *Callinicus: A Defence of Chemical Warfare*, London: E.P. Dutton & Co. (1925).

Hanslian, Rudolf, *Der Chemische Krieg*, Berlin: E.S. Mittler und Sohn (1927).

——, *Der Deutsche Gasangriff bet Ypern am 22 April 1915*, Berlin: Luftschutz (1934).

Hardy, Phil, *The Encyclopaedia of Science Fiction Movies*, London: Aurum Press (1984).

Hasse-Lampe, Wilhelm, *Handbuch für das Grubenrettungswesen* (International), Bande 1, Dräger: Lübeck (1924).

Hersh, Seymour, *Chemical and Biological Warfare: America's Hidden Arsenal*, New York: Bobbs-Merrill (1968).

Hoffman, Bruce, *Inside Terrorism*, New York: Columbia University Press (1998).

The Holy Bible.

Irvine, David, *The Destruction of Dresden*, London: Arnold (1966).

——, *Hitler's War*, London: Arnold (1977).

Irwin, W., *The Next War: An Appeal to Common Sense*, New York: Boni & Gaer (1921).

Jensen-Stevenson, Monica, *Kiss the Boys Goodbye*, London: E.P. Dutton & Co. (1990).

Joachimsthaler, Anton, *The Last Days of Hitler: The Legends – The Evidence – The Truth*, London: Arms & Armour Press (1996).

Jones, H., *The War in the Air*, 5 Vols, Oxford: Oxford University Press (1935).

Kahn, M.K., *CBW in Use: Vietnam*, London: George G. Harrap (1968).

Kaplan, David E. and Marshall, Andrew, *The Cult at the End of the World*, New York: Crown (1996).

Kleber, B.E. and Birdsall, D., *The Chemical Warfare Service: Chemicals in Combat*, Washington DC: The Chemical Warfare Service (1966).

Klotz, H., *Germany's Secret Armaments*, London: Arnold (1934).

Lacquer, Walter, *The New Terrorism: Fanaticism and the Arms of Mass Destruction*, Oxford: Oxford University Press (1999).

Laing, Stuart, *Hitler Diary 1917–1945*, London: Marshall Cavendish (1980).

Lefebure, Victor, *The Riddle of the Rhine: Chemical Strategy in Peace and War*, London: The Chemical Foundation (1919).

Liepmann, H., *Death from the Skies: A Study of Gas and Microbiological Warfare*, London: Arnold (1937).

Lilienthal, D.E., *The Atomic Energy Years 1945–1950*, New York: Boni & Gaer (1964).

Lohs, Karl Heinz, *Synthetic Poisons: Chemistry, Effects and Military Significance*, 3rd edn, East German Military Publishing House (1963). Published in the United States as JPRS 23.681, Department of Commerce (1964).

Lovell, S.P., *Of Spies and Stratagems*, New York: Boni & Gaer (1963).

Ludendorff, Eric von, *War Memories*, Berlin: E.S. Mittler und Sohn (1919).

The Merck Index of Chemicals and Drugs, 7th edn, New Jersey: Merck & Co., Rahway (1960).

Miller-Kiel, Ulrich, *Die Chemischer Waffe im Weltkrieg und Jetzt*, Berlin: Verlag Chemie (1932).

Noel-Baker, P., *Disarmament*, London: Arnold (1926).

Oschsner, H., *History of German Chemical Warfare in World War II*, Historical Office of the Chief of the US Chemical Corps (1949).

Paxman, J. and Harris, R., *A Higher Form of Killing: The Secret Story of Chemical Warfare*, New York: Hill & Wang (1982).

The Penkovsky Papers, London: n.p. (1965).

Perry-Robinson, Julian, *The CWC: The Success of Chemical Disarmament Negotiations*, SIPRI Yearbook 1993. Oxford: Oxford University Press (1993).

——, *World Health Organisation Guidance*, Geneva: World Health Organisation (2004).

Pershing, J.J., *Final Report of General John J. Pershing*, Washington DC: US Government Printing Office (1920).

Prentiss, A.M., *Chemicals in War: A Treatise on Chemical Warfare*, New York: McGraw-Hill (1937).

Preston, Richard, *The Cobra Event*, New York: Random House (1998).

Quester, G.H., *Deterrence Before Hiroshima*, New York: E.P. Dutton & Co (1966).

Roberts, Brad (ed.), *Terrorism and Chemical and Biological Weapons*, Arlington, VA: The Chemical and Biological Arms Control Institute (1997).

Roberts, David, *Minds At War*, West Sussex: Saxon Books (1996).

Rose, Stephen (ed.), *Chemical and Biological Warfare*, London: George G. Harrap (1968).

Schwarte, H., *Die Technik Im Weltkrieg*, Berlin: E.S. Mittler und Sohn (1920).

Simon, Jeffrey D., *Toxic Terror: Assessing the Terrorist Use of Chemical and Biological Weapons*, Cambridge: MIT Press (2000).

Smithson, Amy E., *Separating Fact from Fiction: The Australia Group and the Chemical Warfare Convention*, Washington DC: Henry L. Stimson Center (1997).

——, *Ataxia: The Chemical and Biological Threat and the United States Response*, Washington DC: Henry L. Stimson Center (1999).
Speer, Albert, *Inside the Third Reich*, London: Phoenix (1995).
Stepanov, A.A., *Chemical Weapons and Principles of Antichemical Defence*, Moscow: n.p. (1962).
Stockholm International Peace Research Institute (SIPRI), *The Problem of Chemical and Biological Warfare*, Vols I–VI, Stockholm: Stockholm International Peace Research Institute (1971–1975).
Taylor, Frederick, *Dresden Tuesday 13 February 1945*, London: Bloomsbury (2004).
Thuiller, H.F., *Gas in the Next War*, London: Geoffrey Bles (1939).
Trevor-Roper, H.R., *The Last Days of Hitler*, London: Penguin Books (1962).
Trial of Major War Criminals, 43 Vols, Nuremberg International Tribunal (1947).
Verweg, Wil D., *Riot Control Agents and Herbicides in War*, Leyden, the Netherlands: A.W. Sijthoff International Publishing Company BV (1977).
Waitt, Alden H., *Gas Warfare and the Chemical Weapon: Its Use and Protection Against It*, New York: Duell, Sloane & Pearce (1942).
Watkins, T., Cackett, J.C. and Hall, R.G., *Chemicals, Pyrotechnics and the Fireworks Industry*, Oxford: Permagon Press (1968).

Newspapers and Magazines

Baltimore Sun (12 November 1941; 15 January 1967).
Daily Telegraph (8 July 1963; 6–8 June 1967; 23 July 1968; 23 April 2003).
Financial Times (18 July 1967).
Glasgow Sunday Herald, 'Britains Chemical Bazaar', Neil MacKay (9 June 2002).
Japan Times (4 October 1970).
'Understanding the Threat', *Kuwait Samacher* (2003).
L'Humanité (20 December 2000).
Newsweek (4 March 1991).
The New York Times (21 February 1938; 23 March 1955; 24 March 1965; 25 March 1965; 26 March 1965; 13 January 1966; 10 September 1966; 6–8 June 1967; 13 December 1995; 18 December 1995; 23 April 1997; 28 April 1997).
New York Tribune (26 April 1915; 27 April 1915).
Pittsburgh Post-Gazette (4 August 1995).
Scandinavian Times (November 1970).
St Louis Post-Dispatch (11 July 1997).
Sunday Herald (9 February 2003).
Sunday Times, 'Hitler's Deadly Secrets' (22 February 1981); 'After 24 years – dumped war gas hits holiday beaches' (10 August 1969).
The Economist, 'The New Terrorism – Coming Soon to a City Near You' (15 August 1998).
The Guardian (6 March 2003).
The Independent (7 July 2004).
The Times (6 September 1939; 13 October 1939; 21 November 1939; 11 May 1942; 10 March 1960; 17 May 1965; 6–8 June 1967; 11 December 1967; 23 October 1996).
Time Magazine (2 April 1965; 15 June 1998).
USA Today (4 December 1978; 28 January 1991).
Washington Post (18 October 1966; 6 June 1967; 7–8 June 1967; 19 March 2000; 31 March 2000).
Washington Times (15 August 1992).

Websites

Note: Readers should be aware of the transience of Internet links. All those listed here were alive in 2004.

www.ccfr.org/publications/opinion/opinion.html
www.cia.gov
www.cnn/news/archive.htm
www.copvicia.com
www.ddh.nl
www.fas.org/irp/threat/an253stc.htm
www.geocities.com/rollofhonor32/downingst.html
www.globalsecurity.org/wmd/library/1984/ARW.htm
www.guardian.ch.uk/3858.4571480.htm
www.gwu.edu/~nsarchiv/NSAEBB
www.imperialservices.org.co/military_tactics.html
www.int.org/chem.war0503a.html
www.kcl.ac.uk/Lncma/pro/p-pre08.htm
www.mod.uk/issues/gulfwar/info/ukchemical.htm
www.munition.gov.ru/eng/oldobijil.htm
www.nobelprizes/1919.htm
www.nobel.se/laureates/lecture.html
www.osins.sunderland.ac.uk/autism/hooper2000htm
www.pm.gov.uk
www.rand.org/terrorism.htm
www.stpetersburg/text/htm
www.teamhouse.tni.net/Tailwind/Ruppert.html
www.thehistorian.co.uk
www.tripod.com/mapchemxxx.htm
www.users.globalnet.co.uk/dccfarr/gas.htm
www.vectorsite.net/gas.html
www.vectorsite.net/gas2.html
www.vectorsite.net/gasTW5.html
www.vnh.orh/Chapter2.htm
www.washingtonpost.com/ac2/wp-dyn.htm
www.un.org/Depts/Unmovic?Bx27.htm
www.znet.org (The Pilger Reports)

Interviews

Tom Jackson (1 November 2003).
Julian Perry Robinson, SPRU, University of Sussex (5 July 2004).

Theses/papers

Coleman, Kim, 'IG Farbenindustrie AG and Imperial Chemical Industries; Strategies for Growth and Survival 1926–1953', PhD thesis, University of London (2001).
Ganor, Boaz, 'Israel's Counter-Terrorism Strategy', PhD in progress, Hebrew University (2003).

Heller, Major Charles E., USAR, Chemical Warfare in World War I: The American Experience, 1917–1918, Combined Arms Research Library, Fort Leavenworth, Kansas, Leavenworth Paper No. 10.

McLeish, Caitriona, 'The Governance of Dual-Use Technologies in Chemical Warfare', MSc thesis, SPRU, University of Sussex (1997).

Swyter, H., Proceedings of a Conference on Chemical and Biological Warfare, Salk Institute, Brookline (25 July 1969).

Tu, Anthony, 'Chemistry and Toxicology of Nerve Gas Incidents in Japan in 1994 and 1995', paper presented to the 3rd Congress of Toxicology in Developing Countries, Cairo (1996).

Television documentaries

PBS, *Race for the Superbomb* (January 1999).
BBC, Horizon, *Gulf War Jigsaw* (14 May 1998).

Diaries

Lance-Corporal A. Barclay, 1/1st Ayrshire Yeomanry (1914–1918).

Index